Writing the Liberal Arts and Sciences

Writing the Liberal Arts and Sciences

*Truth, Dialogue, and
Historical Consciousness*

*Edited by
Mary Bouquet,
Annemieke Meijer and
Cornelus Sanders*

Amsterdam University Press

Cover illustration: Floor Plan University College Utrecht
Reproduced with permission of Architectuurbureau Sluijmer en van Leeuwen BV

Cover design: Gijs Mathijs Ontwerpers, Amsterdam
Typesetting: Crius Group, Hulshout

ISBN   978 94 6372 936 9
e-ISBN   978 90 4855 508 6 (ePub)
DOI   10.5117/ 9789463729369
NUR   113

Printed and bound by CPI Group (UK) Ltd, Croydon, CR0 4YY

Table of Contents

III   Historical Consciousness

Acknowledgements

We gratefully acknowledge all those who presented papers at the Scholarly Stories seminar, 2017-2020, helping to create a fruitful setting for exchange among liberal arts and sciences practitioners at our Tuesday lunchtime meetings in Locke A. These include Sandra Ponzanesi, Bettina Boschen, Erwin van Sas, Robert Renes, Lonia Jakubowska (who initiated the series), Meindert Fennema, Wil Pansters, Sabine Uijl, Jan van Ophuijsen, Maarten Eppinga, Kiene Brillenburg Wurth, Diana Odier-Contreras Garduno, Christoph Baumgartner, Gert Jan Vroege, Nina Köll, Brian Dermody, Patricia Canning, Bas Defize, Gerda van Rossum, Arjen Vredenburg (and cello), Tatiana Bruni, Chiel Kattenbelt, Kim van der Wijngaart, Guido Terra-Bleeker, Maarten Diederix, Mark Baldwin, Elisa Veini, Ingrid Wesselink, Konstantina Georgelou. Even though time and other constraints conspired against their writing an essay for our rather open-ended project, we nonetheless benefited from all of these contributions for cultivating a liberal arts and sciences habitus in our community.

Our heartfelt thanks to Inge van der Bijl of Amsterdam University Press for graciously receiving us and our proposal for a book and encouraging us at every step, and to AUP Desk Editor Jaap Wagenaar for his cheerful efficiency.

Former Dean of UCU, James Kennedy, gave the project his blessing, encouraged, and supported the full-day seminar we ran at Antropia, Driebergen; we greatly appreciate his trust and participation. Without the collegial understanding and support of Sjoerd Bosgra at decisive moments in the book's coming-into-being, this project risked turning into a damp squib: sincere thanks for steady nerves. We also thank Thomas Hardeman for his incisive work on the index and Peter de Voogd for taking in hand our endnotes. We are grateful to Michael van Leeuwen for permission to use his 'Masterplan Village Campus' for the cover of this book.

We have also greatly enjoyed and learned from one another in creating this book: bringing together the different perspectives and competences that go into writing and editing a book has been a non-stop learning process.

Mary Bouquet, Annemieke Meijer and Cornelus Sanders
Utrecht, 30 April 2021

Introduction: Writing the Liberal Arts and Sciences

Mary Bouquet, Annemieke Meijer and Cornelus Sanders

This book was written by authors from the fields of literary studies, criminology, physics, linguistics, political science, medicine, philosophy, clinical psychology, mathematics, art history, law, astrophysics, history, anthropology, and religious studies. It is the written outcome of a form of collegial storytelling among faculty members working in the liberal arts and sciences setting of University College Utrecht, the Netherlands. Weekly lunchtime gatherings brought together academics from these widely differing fields, with equally diverse personal journeys, to sample other fields of knowledge. There was a combined sense of curiosity about the content of these other fields and those in them; about different ways of approaching and understanding the world; and about the possibility of being able to ask questions outside one's comfort zone. This interest was fostered by working together in a multidisciplinary liberal arts and sciences college; sharing offices and other facilities with colleagues from completely different fields meant that we were used to having chats across disciplinary boundaries, probably more so than our colleagues in monodisciplinary departments. We wanted to see where these conversations would lead if channelled into written form. The willingness to think beyond the boundaries of a given field when approaching a particular phenomenon proved stimulating as a writing experiment: critical feedback on content and style from a diverse body of faculty members was unprecedented. We began to wonder whether we could write about some aspect of our respective fields of knowledge not just for ourselves but also for external audiences. We imagined our readers as potential liberal arts and sciences students, but also as interested peers and lay people. As we proceeded, the subjects we wrote about changed and evolved. The many-faceted process of transformation from the spoken to the written word reflects the effort of moving beyond disciplinary jargon. The interdisciplinary in this respect is a particular kind of scholarly attitude built upon curiosity and open-endedness: we did not know the outcome in advance, but allowed our voices to emerge *through* a process that stretched over several years.

The themes of truth, dialogue, and historical consciousness surfaced slowly – while contributors moved from storytelling mode to the written essay format – and became a possible way of sequencing the texts. That said, each essay exceeds the category to which it was assigned, intersecting with the other two. A sense of history surely informs the quest for truth through language in the realm of philosophy. Truth can also be revealed through the language of fiction: novels, poems, and plays, written at various historical junctures, use various conventions that establish truths in fictional worlds that are nonetheless part of the world. Trials involve the deployment of evidence and argument by prosecution and defence, calling upon witnesses, requiring a jury to arrive at a truthful verdict, and presided over by a judge. Loaded questions that plant ideas in people's minds are a means of mystifying history and obscuring the truth by means of syntax. Language, as a dynamic interface between cognitive and environmental realities, each of which translates and shapes the other, entails that some meanings exceed grammatical relations. The material forms of sacred texts through history exercise particular kinds of agency through their liturgical and ceremonial uses, combining the Word with print and digital technologies that call into question assumptions about secularization and modernity.

Notions of the truth are conceptualized and approached quite differently in the respective domains of philosophy, literature, criminology, linguistics, and religious studies, due to their respective histories. Yet the contributions from these disciplines show that approach and the subject matter are not simply 'different', but that questions, conversations, and dialogue across those disciplinary boundaries opened up new perspectives and deepened the awareness of the position of each domain of knowledge with regard to the others. Growing awareness of the fragility of our world, and the damage inflicted by collective consumption patterns and engrained assumptions, has infiltrated many disciplines and was sharpened by the global Covid-19 pandemic that struck in 2019/2020. This biosocial crisis gave an impetus to our writing: writing *as* a form of social life, a search for truth and dialogue, when our professional lives took an unexpected turn: the transformation of the university into a digitalized cottage industry brought home the core value of writing projects. Amid the uncertainty wrought by a virus, writing became a way of engaging with and reflecting upon the world, a way of adding value to the fragmented world we found ourselves in. Writing became, quite literally, a moment of truth and sustenance at a moment of historical rupture.

Authors have recourse to several literary forms including dialogue, poetry, fiction, and the written assignment. These forms provide means of reconnoitring such matters as the archaic language of seventeenth-century philosophers, the eloquent silence of monuments, and the emergence of a voice by means of a college assignment. An imagined dialogue between non-contemporary Western- and Eastern-schooled philosophers opens up new vistas on the central question of being in the world. Writing poetry offers law students a new entrée for reading political philosophy and making it their own. Case studies written as human stories offer new insights into the decisive conversations between medical practitioners and their patients which usually take place behind closed doors. Taking account of local healing systems in low-income countries by engaging trained paraprofessionals helps to bridge the gap in mental health care while changing the course of clinical psychology. The letters exchanged between historical scientists, who were also friends, elucidates how the broader intellectual setting can enrich or impoverish the development of knowledge in a particular field. Literalizing a metaphorical dialogue between two monuments is a way of exploring the agency of aesthetics, space, materials, and form, which can reinforce understanding while going beyond words.

The essays exemplify historical consciousness in equally varied ways. These include writing about paradigm shifts within specific disciplines, which is a classic way of introducing and contextualizing a more specific topic. The historical shift from quantifying concrete phenomena to characterizing and quantifying abstract categories, in the field of statistics, aptly illustrates this point. A related point concerns the impact of an increasing body of knowledge within a particular field: mathematics. Since it takes more of the student's lifetime to gain cognizance of the field of mathematics, the result is increasing specialization and the loss of a holistic vision of knowledge. Consciousness of the role of history in the present also impels the updating of national historical canons – where these exist. History is reconfigured when government calls upon specialists from the next generation to update the events, protagonists and phenomena featured in the canon in the light of current social issues and developments. This top-down version of history contrasts with the grassroots contestation and revision of a government-sanctioned version of the history of the energy sector by means of public art/ monuments. While visual representations of the self, led by the selfie, are ubiquitous, few are aware of the historical conditioning of the pose they adopt, the attire they don, and the attributes with which they surround themselves. The truth is that such portraits have been a matter of debate and censure for centuries: historical awareness here functions almost as a cautionary tale. The photographic portrait of the black hole at the other end of the universe, an event that happened lightyears ago, mesmerized public attention. The image was an outcome of dialogue and cooperation in the global scientific community of astrophysicists, searching for something that is known but invisible. Faced with this sense of universal historical catastrophe, the practice of writing, of finding a voice, gains new élan.

Dear Reader, we wish you much pleasure in perusing our essays – in any order you wish – which we hope may inspire you to explore for yourself the joys of writing, and of the Liberal Arts.

I

Truth

..., *specif* : a convi
privileges

truth \\'trüth\\ *n, pl* **truth.**
fidelity; akin to OE trē
archaic : FIDELITY, CON
utterance 2 a (1) : t
of real things, events, ar
dent fundamental or sp
idea that ...

The indispensable Truth: Postmodernism and the possibility to understand each other

Floris van der Burg

> The ideal subject of totalitarian rule is not the convinced Nazi or the convinced Communist, but people for whom the distinction between fact and fiction (i.e. the reality of experience) and the distinction between true and false (i.e. the standards of thought) no longer exist.
> *– Hannah Arendt*[1]

Truth is under siege. Fake news is everywhere and serious research that warns of current climate change is happily impugned by non-experts. The internet is rife with conspiracy theories ranging from the usual alien abductions, to claims about a flat earth, or the dangers of vaccination. We grow up with politicians who have, to say the least, a strained relationship with the truth. Disagreements about the possible truth of a story or position frequently end in: 'Well, that is *my* truth!' or 'That is just your *opinion.*' This is a widespread approach to the confusing times we live in. Every day we are confronted with an enormous amount of information that may or may not be true. How do we decide what to believe? Or does it simply not matter? Is *my* truth indeed as good as any other? Is truth relative to me, or my circumstances, or to a particular culture?

In the wake of the Enlightenment, Modernism (roughly from the seventeenth to the nineteenth century) taught us that logic and reason are universal, and that the discoveries of the natural sciences could, ultimately, be substantiated by irrefutable evidence. That is, the claims of the sciences about the way the world is are objectively true or false. While science yields an ever more truthful picture of the way the world is, this ever-improving understanding of the world we live in will also ensure that human beings will change themselves and their societies for the better. The expectation is that life in the future will be more humane, just, and prosperous.

But not everybody is so optimistic. The twentieth-century movement of Postmodernism, although quite diverse within itself, rejects this Modernist optimism and its faith in Truth, reason, and logic (where 'Truth' with a capital 'T' stands for eternal, unchanging truth). Characterized by a deep suspicion of any claims to universality, the Postmodernists often deny the very existence of objective truth, and the universal validity of reason and logic. Truth, some argue, is relative to a culture, and any attempt to impose one's own truth on another culture might even be considered an act of aggression, an attempt to replace their truth with yours.

Such scepticism about the very possibility of our sentences, theories, or beliefs accurately describing an objective world has taken hold of many of us, quite independently of any knowledge of, or even an interest in, Postmodernism. There are some compelling reasons for this: widespread disagreement about the nature of many aspects of our lives is common, and, rightly, often encouraged. We disagree about politics, morality, philosophy, and aesthetics. Even in scientific debate we disagree about how best to understand the world. The history of science is not a neatly progressive story of humankind slowly encroaching upon the final, True story of nature. Instead, it is the story of a messy, haphazard series of hypotheses, theories, and arguments between scientists. Could it be that this disagreement is merely symptomatic of a deeper problem: the absence of an objective, eternal Truth?

As our view of the world expands and we are increasingly aware of the different points of view held by different people on many topics all over the globe, we are ever strengthened in our scepticism about claims to Truth. Such relativism, especially about issues to do with values, is not new. It simply comes with this wider view of the world that includes more, different cultures. It is unsurprising then, that the relativism of Postmodernism fell on fertile ground. By arguing against the universality of our general philosophical, scientific, and moral assumptions, the Postmodernists pointed to the boundaries of the validity of *our*, often Western, ways of thinking. Perhaps the world is *not* the same for everyone, perhaps there are alternative versions of logic, perhaps there is not only one single human nature, and perhaps science is relative to culture too.

It may be possible to explain the popularity of the Postmodernist movement by the many amusing and creative applications of its very basic tenets. If science is not a collection of accurate descriptions of the way the world really is, then it might just be a story, a story like any other story, not better and not worse. Inside the story or narrative one would be tempted to accept its truth as *the* Truth, but from the outside this may not be so obvious. Stories then, are ideal to illustrate the point of Postmodern thinking. We often hear Coleridge's phrase 'suspension of disbelief'[2] when we accept the truth or plausibility of certain ingredients in a story which we would reject in the real world. Think of the magic in the Harry Potter stories, or the uncanny ability of the spaceships in *Star Trek* to fly at multiple times the speed of light. In such cases the boundaries between contexts, or truth-worlds, are clearly drawn. But a more powerful way to raise awareness of different truth-worlds may be to deliberately cross the boundaries between them.

Consider this example. In the 1986 John Hughes film *Ferris Bueller's Day Off* we see the lead character, Ferris Bueller, repeatedly turn away from the action in the movie and address the audience in the cinema about the content of the film.[3] Famously, he tells the audience about his reasons for taking the day off school: 'Life moves pretty fast. If you don't stop and look around once in a while, you could miss it.' In the final scene, when the credits are already rolling, Ferris returns on screen and says: 'You're still here? It's over. Go home.' Ferris Bueller undoubtedly was not the first one to step outside the context of the work of art he was part of, while staying in character, and he will not be the last. The Postmodern point of such a context-busting device is to deliberately cross the boundary between two different contexts, between two different truth-worlds. Truth, in such contexts, is relative to its truth-world. A truth-world can be a culture, a system of values, a scientific theory, or, as in our example, a movie. Ferris Bueller's stepping out of the movie to address the audience in the cinema confronts us with the relative nature of all that is said within the context of a single truth-world. It draws attention to the boundary between our world and the movie-world, where things are different, where different statements are true and false.

If truth can depend on the particular narrative it is part of, be it a story, a theory, a system of values, or a culture, it is no wonder many people are suspicious of any and all truth-claims. Why should we believe any of these claims? Why should we regard truth as any more stable than a 'mere' opinion? As said, then, truth, and especially *Truth*, has earned itself a bad reputation.

It may well be fair and reasonable to question the truth of the content of the claims that we make about politics, ethics, aesthetics, and, yes, even science. But should we doubt the validity of these claims just *because* they are truth-claims? The truth or truth-claims is what we can doubt. There is good reason to do so on many occasions. The best reason for a critical attitude towards truth-claims is that we do not know that these truth-claims are indeed the Truth (note the capital T). Thus Truth is elusive. Some thinkers, known as radical sceptics, even say that we can never attain Truth, not even in the best and most fruitful sciences. They have good cause to say this: it is not possible to find conclusive confirmation by means of experiment of *any* theory. This is not some mysterious rule, but rather a simple consequence of what our theories (laws) in experimental science do. They govern events of a certain type in the past, the present, *and* the future. The problem is, of course, that we cannot confirm by experiment that our theory or law will always work in the future, no matter how successful it has been in the past. So there always remains an element of doubt.

When we use the word 'true' for less exact, or better, less exceptionless areas of study, such as ethics, aesthetics, or politics, then there appears to be even more reason to question such truth-claims. It is difficult enough to try and formulate strict laws that predict the behaviour of the lifeless world of the material objects that surround us. But when the behaviour we are trying to capture in universal generalizations is the behaviour of human beings, then our inclination to doubt the universality of these laws is even stronger. Perhaps there is a simple explanation for this tendency: we simply do not like it when we are explained by means of law-like regularities. It does not do justice to our conviction that we are individuals whose behaviour warrants a complex, individual explanation. Our strong sense of individuality rebels against any temptation of the natural sciences to say that our behaviour, our innermost motives and desires, may well be reduced to the behaviour of perfectly predictable physical particles. This type of physicalist reductionism does not fit in well with the unique value we like to ascribe to every human individual, or as the l'Oréal ads like to tell us: 'You're worth it!' You are unique and exceptional.

It is entirely in line with this powerful sense of self then, that we are strongly inclined to reject all truth-claims that are not our own. If, by the lights of some Postmodernists, it may be the case that claiming a truth that transcends the individual, is at once an attempt to replace another culture's truth with yours, then *any* truth-claim may also be considered an assault on *your* personal truth. Cultural relativism, in these cases, easily descends into individual relativism.

Fortunately, this strikes most of us as an undesirable situation, because it is unworkable. We resolve it, in most cases, by separating the individual, or even cultural, claim from a more universal one. So, we tend to allow the outcomes of thorough research on the efficacy of a certain painkiller to carry more universality than claims about the beauty of the Mona Lisa. 'Beauty is in the eye of the beholder', we say, but the power of morphine is not. But some Postmodern thinkers insist that *all* truth is relative to a culture.[4] This would then render even the most well-established claims of the empirical sciences non-universal. It is quite easy to maintain such a stance as there are many sensible scientists who are radical sceptics. So, they would not claim absolute truth for any theory. It is not difficult to see how this cautious attitude towards science can be used to impugn all truth claims, as there always remains an element of doubt about what we appear to know to be true. And yet there is good reason not to give in to this temptation.

How then should we respond to such varied and widespread denial of the possibility of objective truth? The most important difference between a relativist and a radical sceptic is that the relativist denies that there is Truth. That is, the relativist simply says that there is no one way the world really is. The radical sceptic is perfectly happy to say that there is Truth, that there is one way the world really is, but we never know whether we got it right. This may appear a subtle difference, but it is not. The relativist has given up, whereas the radical sceptic can happily continue to investigate the nature of the world. The sceptic sees scepticism as a methodological stance: it makes good sense to question all your conclusions, even if you are quite certain of them. But when your critical attitude does *not* force you to admit that you are wrong (this time), you may of course hold on to this theory until it is indeed falsified by new insights or evidence. This way we do not need to give up on all our sciences as 'not True anyway', or 'just another opinion'. The crucial difference here is the difference between the expertise of a specialist in a certain field and the lay opinions of the non-specialist in that field.

Still, the dogged relativist will not be persuaded by any of these arguments. Nothing that has been said so far makes it impossible to stick to the claim that there is no Truth, if for no other reason than that it is not possible to establish any of these 'Truths', even by the radical sceptical scientists' own admission. So, even if we have, and I suspect that we indeed have done this, established that there are very good practical reasons to work at finding out the Truth about the world we live in, this will not dissuade the relativist from the position that there is simply no Truth. Perhaps, the relativist will argue, this obsession with Truth is unhealthy, and it would be better for all if we admitted that truth is relative, and hence there is no role for truth left in our (post)modern world. Sometimes it is only a minuscule sliver of doubt that gives the relativist enough reason to impugn even our most reliable theories as 'not True'.

We might respond by giving up on the end-goal of Truth in science, and turn to a form of pragmatism: what works stays, what does not work, goes. And the term 'truth' has become obsolete. I suspect this is not an advisable route to take, if for no other reason than that very often in the history of science, we have given up empirical success (what works) in favour of a new theory that did not work so well, but seemed Truer. What works is not the only criterion when choosing which theory to embrace. Certainly, anyone concerned with the social, political, or power-implications of truth claims, in science and beyond, must be very cautious around purely pragmatic considerations in deciding how to view the world. Unjust, but eminently pragmatic solutions to moral or political problems continue to offend our most fundamental sensibilities.

But what do we say to the relativist, who is principled, and still wants to give up on all uses of 'truth'? We must say that the relativist cannot dispose of truth. The reason is as simple as it is far-reaching. Truth is not part of the world. Truth is a function of language. What can be true or false are sentences claiming something. What is true or false are theories, propositions, statements, or stories. So when our relativist mistakenly identifies Truth with 'the World as it really is', she is confusing the world being described in our sentences and theories with the predicate 'is True' that we use to class these sentences and theories. The world is not true or false; the world just *is*. That is, trees and dogs and fire hydrants and brick buildings are not true or false, but the sentences we use to talk about them are. It is here that the *concept of truth* becomes relevant. The relativist may be able to maintain that no sentence is reliably True, but that does not mean that the relativist can also claim that we have no further use for the concept of truth. This concept is not part of the world. It is part of language. If we did not have the concept of truth in our language, we could not understand each other. If we did not have the concept of truth the relativist could not even *begin* to speak of theories that are all just stories, the one no better or worse than the other.

An example: if I say 'I believe the sun is shining', what should my audience then assume about me? Surely it is that I believe the sentence 'the sun is shining' is true in the context in which I uttered it. Imagine a world in which we do *not* think that when people say what they believe, they also think that their belief is true. In this world our language would no longer make sense. And we could no longer make sense of what people are trying to tell us. This use of the concept of truth cannot be abolished by the relativist, on pain of abolishing all language and with it the wonderful literature of which the Postmodern relativist is so fond.

If this is right, and I believe that it is, then the relativist is not much more than the radical sceptic, who believes that there is always some remaining question as to the Truth of any theory. For the relativist this scepticism may then also include all beliefs and all narratives. But again, that is no different from the radical sceptic, who simply never even considered quibbling over the truth of the magic in the Harry Potter novels. That the truth or falsehood of such linguistic expressions is context-bound was long clear to the sceptic. Indeed, the degree to which we can or should doubt the truth of some story, theory, or statement is *relative* to the domain to which it belongs. But that says something about the different domains, such as narrative, art, politics, science, or maths, to name but a few, and not about truth itself.

Suggestions for further reading

C. Butler. *Postmodernism: A Very Short Introduction.* Oxford: Oxford University Press, 2002.

R.L. Kirkham. *Theories of Truth: A Critical Introduction.* Cambridge, MA: MIT Press, 1995.

S. Read. *Thinking About Logic: An Introduction to The Philosophy of Logic.* Oxford: Oxford University Press, 1995.

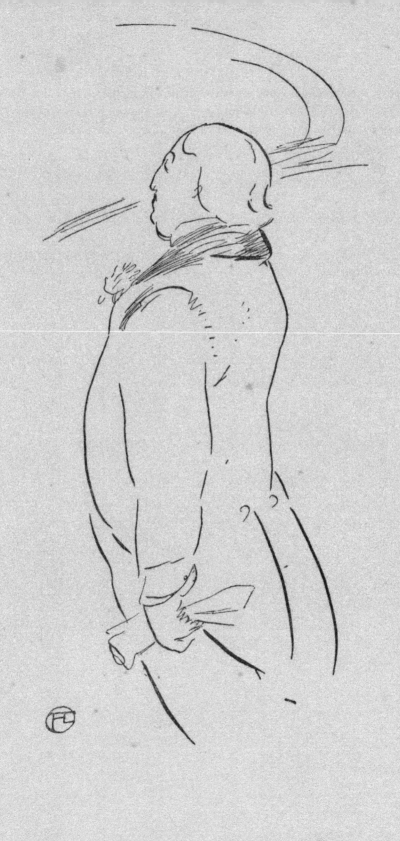

Fictionality, or the importance of being earnest

Agnes Andeweg

If I were to ask you to give some examples of fiction, chances are you would come up with a book or film with an invented story, such as *Star Wars*, or *The Fault in Our Stars*. Fiction, so the common definition goes, contains imaginary characters or events. But in spite of that seemingly straightforward definition, things quickly become more complicated once we start thinking them through. As we can tell from just these two examples, fictional texts come in various shapes and gradations of invention. John Green's bestseller *The Fault in Our Stars* (2012), for example, features an imaginary writer (Peter van Houten) and his imaginary work, but also refers to historical writers (Shakespeare), and it is set in cities that exist in reality (Indianapolis, Amsterdam). *Star Wars*, on the other hand, takes place in an imaginary galaxy peopled by both human and non-human characters. On imaginary planets like Tatooine and Hoth many things happen that are simply impossible on Earth.

So, what then is fiction, exactly, and how do we recognize it? Are there any features all fictional texts share? And should we even limit the search for characteristics of fictionality to texts? These complex questions have been discussed for many decades by, among others, literary theorists, philosophers, and psychologists. To dip into the latter question: while we most commonly associate fiction with a type of texts, scholars have argued that it can also be found beyond the page if we realize fiction can be considered a form of make-believe, a game of pretence. This expands the realm of fiction to the situation where a parent cries, 'Here comes the monster that is going to eat you!' and makes their child scream with laughter. The child is clearly able to recognize the make-believe, or it would immediately flee. Or think of children's games where a tree becomes a house, or snippets of paper are turned into money. Children act out made-up situations like 'You were the patient, and I was going to make you better'. Interestingly, these are often enacted in the past tense, which points to the close proximity of fictionality and narrativity: where there is fiction, there are stories, so it seems.[1]

Speaking of pieces of paper: is not our use of money one big game of make-believe, a pretension that allows us to do real transactions by using digital numbers and pieces of paper? Isn't this fiction, too? Well, to an extent: just like in children's games, we can make something stand in for something else by using a prop. This process of representation – to make something stand in for something else – seems crucial for fictionality. But using money is also different from a children's game. When we use the prop of money, this use is bound by very strict rules: I cannot use just any sort of paper to pay you. The purpose of using money is to convey something in our actual world, rather than to imagine a possible world which we are free to invent. Arguably, encouraging the imagination is also a condition for fiction.

Lawyers know fiction as well: law has a so-called *fictio legis*. This legal fiction is a situation which is known to be untrue but nevertheless taken for a fact by the court. For example, in Roman law only free citizens were allowed to make a testament. To prevent the situation that a citizen who was imprisoned would not be able to make a testament, a law was made in which it was assumed that Roman citizens in prison were still free. Here the fiction does not so much create a new possible world, as *Star Wars* does, but rather turns a non-existing situation (the free Roman) into the actual one. Thus, the *fictio legis* 'fixes' a blatant discrepancy between the actual world and the desired reality.

Another field where we – perhaps surprisingly – may find a type of fiction is mathematics, which has imaginary numbers. An imaginary number is a real number (like 1, -6, π or $\sqrt{2}$) combined with an imaginary unit which has -1 as its square ($i^2 = -1$). We all learnt in school that squares cannot be negative, so does that make imaginary numbers fictions? They are surely representations, props if you will, of a non-actual world, but that could be said about a lot of mathematics. And do imaginary numbers encourage playing with the imagination? Probably they do, though the purpose may not be playing in itself but solving serious mathematical problems. But perhaps we should not separate play and seriousness so strictly but rather stress the common factor of creativity. What would any scientist be without the imagination, without the capacity to speculate about possible worlds? Their fictions may become reality.

Apart from lawyers and mathematicians, philosophers have wrestled with the issue of fiction and truth. What toddlers grasp intuitively – that in stories ducks can talk and elephants can fly – has turned out to be a stubborn philosophical problem. How come we agree that the statement 'A unicorn has one horn' is true, and 'A unicorn has two horns' is not, when unicorns do not exist in the first place? How can we make truth statements about fictional entities? Humans use language to state things about the world, but apparently, we have to assume there are fictional worlds which have their own logic, and their own truths.

The interest of cognitive neuroscientists in fiction stems from a similar wonder as the philosophers'. How is it possible that readers can be moved to tears by events they are perfectly aware of never happened, like the death of Albus Dumbledore? What does reading fiction do to your brain? Clearly, we are able to respond emotionally to fictional situations just as strongly as to actual ones. Neuroscientists have also shown that 'flowery' figurative language, such as the use of metaphors, produces a different response than literal language. Language that is less directly representational (a more imaginative 'prop' so to say) triggers a different part of the brain.

This brings us back to the question of language: how do we recognize fiction? Are there any features, or 'signposts' that mark a text as fictional? Literary scholars have long debated this question, but let us try to work out a first answer using the following text:

Mrs Darwin

7 April 1852

Went to the Zoo
I said to Him—
Something about that Chimpanzee over there reminds me of you.

It is likely that many readers will recognize some features of this text as belonging to the genre of the diary: the date at the top, and the ellipted sentence 'Went to the zoo' are typical for a diary entry. Without any information about the author or the context in which this text appeared, these features are signposts of factuality rather than of fictionality. People write diaries to report or reflect on their lives, perhaps being more honest than in any other type of self-representation, rather than to make something up. Also, the name Mrs Darwin combined with the date seems to tie the text to the real world: even if you do not know whether Charles Darwin was married, you will infer it from this text and be inclined to take it for a fact, especially when you have identified the genre as a diary. Readers who happen to know a little bit about evolution theory may even be tempted to construct a chronology of events: when was Darwin's *The Origin of Species* published? (In 1859). This also signals a desire to anchor the event described in the text in a historical reality.

So far, the signs point to factuality, but how about fictionality? Are there any signals that this text is *not* an actual diary entry? Here many readers

will point at the rhyme. Rhyme distracts from content, it is language that
draws attention to itself *as language*, something we do not usually find in
texts that try to convey a message as transparently as possible. Rhyme can
only be noticed in hindsight, at the repetition of a sound, so you probably
noticed the rhyme of 'Zoo' and 'you' only when you reached the last word of
the text – which suddenly turned this presumed diary entry into a poem. It
is also only in hindsight then that one can notice that the rhyme is already
present in the date (eighteen fifty-two). Do not worry if you did not notice
this, it only confirms we tend to read for content (date) rather than for form
(rhyme) – perhaps even more so with numbers than with words. But once
you have realized that 1852 rhymes with Zoo and you, the date as an anchor
to historical reality is called into question. Or as somebody remarked when
I discussed this text with a group: 'if it had been 1857, she would have had
to go to heaven!'

There is another sign that we are dealing with a fictional text here: the
fact that 'Mrs Darwin' is placed above the diary entry indicates that she is
presented as the narrator of this text by someone else; in other words that
the 'lyrical I' is not the same person as the author. Information about this
author cannot be found inside the text but belongs to what literary scholars
call the *paratext*: everything that surrounds it, from the preface to the
author's name. Indeed, revealing the provenance of this text helps to identify
it as fictional: 'Mrs Darwin' first appeared in a poetry volume *The World's
Wife* (1999), by Carol Ann Duffy. Duffy was the British Poet Laureate from
2009-2019. By now we are pretty sure we are dealing with a fictional text.

As a side note: the only genre to actually use the label 'fiction', apart from
non-fiction, is science fiction, which is usually associated with speculations
about the future – think *Star Wars*. But 'Mrs Darwin' could be characterized
as science fiction as well, even though it is set in the past: Duffy rewrites the
primal scene of evolution theory from an adventurous expedition (Charles
Darwin on the *Beagle*) into a family outing, from bird watching on the exotic
Galapagos to visiting the apes in a domestic zoo. Thus, Duffy suggests a
new, less heroic model of scientific discovery, in which a woman features as
discoverer for a change. Move over Indiana Jones! A rewriting of the history
of science and its dominant cultural scenarios in a mere twenty-three words
is definitely a remarkable achievement.

Besides stylistic devices like rhyme, and a narrator who is not the
author, scholars have suggested other signposts of fictionality, such as
the representation of inner thoughts, extensive use of dialogue, or clearly
non-actual elements (from floating chessboards to zombies). There are
fictional texts, however, which do not contain any signposts and are still

fictions (mockumentaries), and fictional signposts can be found in texts labelled as non-fiction (interviews, autobiographies). So scholars now agree that it is not possible to come up with a watertight definition of fictionality based on textual elements alone. Something else is needed as well, and this does not come from the characteristics of a text or its author. The fact that I can ask you to ponder the question of fictionality implies that you have a conception of what fiction is, or fictionality. You know what to look for, you know what to expect. This implies there is a 'pact' between you and the author, a communication situation in which both partners have agreed on how to understand these texts: as fictional. The British poet Samuel Coleridge famously called this pact 'the willing suspension of disbelief': the reader's willingness to accept, for a while, a world that does not actually exist.

This means that fictionality has at least two aspects: not just *invention*, as the Wikipedia definition states, but also *intention*. The fictivity of characters and events (invention) is only accepted when it is matched with an agreement (intention) between authors and readers to treat the text as fictional. In the past, fiction has often been accused of telling lies, or of making errors. With the two aspects of invention and intention it is possible to explain why fiction is different from telling lies and making errors. A liar has the intention to deceive, so the pact between author and audience is absent: they do not share the same intention, or the same rhetorical situation, as it is called. An error marks a non-factual state of affairs: the representation does not match the reality. But it is only an error relative to the rhetorical situation: if author and reader share the intention of exchanging facts, errors are possible. If author and reader share the fictional pact, errors can also be made, but these relate to the fictional reality. It is an error if Donald Duck suddenly wears pants, but not if he talks.

The shared understanding of fiction between author and reader can only flourish if there is a cultural acceptance of this type of text: there have to be shared practices, institutes, or spaces that give a licence to act out and tell fiction. We know when we enter a theatre that we may expect something fictional on stage (but not in the cloakroom), just like we know it when we take a book from the shelf marked fiction. Architects, light technicians, but also publishers, reviewers, and bookshops help establish this licenced space for fiction.

This distinction between invention and intention helps us to understand what is going on in conspiracy theories about climate change, Covid-19 or 5G-networks. When mainstream media are accused of producing fake news, it looks like they are blamed for saying things that do not represent reality: for making errors. But through the use of generic descriptions like 'lamestream media', such accusations also cast doubt on the communication situation, the 'pact' for factual texts: on the media's intention to represent factual, non-fictive events. Just like the shared understanding between media and readers/audiences is presently under attack, some scholars argue that the fictional pact is waning, or else how did 'based on a true story' become such a recommendation for fiction?

Let me, by way of conclusion, turn to an example of fiction which is a wonderful play on these two pillars of fictionality, invention and intention: *The Importance of Being Earnest* by Oscar Wilde (1854-1900). Wilde was one of the most famous poets and playwrights of his time; nowadays he is also remembered for the libel trial he tragically lost. In 1895, the same year *The Importance of Being Earnest* premiered, Wilde was convicted of sodomy and sentenced to two years of hard labour; soon after his release he died, a broken man, at age 46. *The Importance of Being Earnest*, a story about two young (heterosexual) couples, has remained his most popular comedy. Many of its comical twists revolve around things that are made up. It is through the invention of a fictive Ernest that the two main characters, the young men Jack Worthing and Algernon Moncrieff, manage to live a double life. As Ernest, Jack lives a libertine life in the city, while in the countryside he acts as the decent Jack, courting Gwendolyn. Meanwhile, Algernon uses the guise of Ernest to bid for Cecily's hand while keeping his visits as Algernon to the mysterious Bunbury a secret. While the nature of the men's second life is never detailed, there is a suggestion of homosexuality: the fiction of Ernest allows both to keep up heterosexual appearances by creating a parallel world. This suggestion is reinforced by the shady gender politics of the play. Both women are so naïve they can easily be deceived, and both absurdly insist that they want to marry a man called Ernest: 'There is something in that name that seems to inspire absolute confidence', as Gwendolyn remarks. Had the play's title been *The Importance of Being Ernest*, it would have stressed the importance of the invention being true, of a representation that matches reality. Now that the title reads 'earnest' instead, it rather stresses the importance of a serious intention; the subtitle 'A trivial comedy for serious people' seems to drive the point home. But this stress on intention is of course only clear on paper: when acted out and performed, the title is as ambiguous as ever. The pun on e(a)rnest brings together all the different aspects of fictionality: the invention of Ernest and the moral appeal to be earnest, to have the right intentions, combined with language drawing attention to itself. Ultimately, fictional truth is even stranger than fictional fiction: Jack turns out to be 'really' Ernest: he always thought he was an orphan, but just in time his real identity is discovered. So does that mean Wilde prefers fact over fiction in the end? Unlikely, as the play makes clear. As a baby, Jack was accidentally swapped by his governess for a novel, a heavy Victorian tome. If life and fiction can so easily change places, then anything is possible. Most of all, Wilde seems to invite us to engage in playing the serious game of fiction.

Suggestions for further reading

D. Cohn. 'Signposts of Fictionality: A Narratological Perspective'. *Poetics Today*, 11(4) (1990), 775-804.

S. Coulson, V.T. Lai. 'Editorial: The Metaphorical Brain'. *Frontiers in Human Neuroscience*, January 5, 2016.

C. Craft. 'Alias Bunbury: Desire and Termination in The Importance of Being Earnest'. *Representations*, 31 (1990), 19-46.

M-L. Ryan. 'Fiction, Cognition, and Non-Verbal Media'. Eds. M. Grishakova, M.-L. Ryan, *Intermediality and Storytelling*. Berlin: De Gruyter, 2011, 8-26.

R. Walsh. *The Rhetoric of Fictionality: Narrative Theory and the Idea of Fiction*. Columbus, OH: The Ohio State University Press, 2007.

IVSTITIA

VIOLENTIA

3

Exuitur Feritas, armisq̃ potentius Æquum,
Sub pedibus calcat surgentia fraudibꝰ arma. *sadeler f.*

And Justice for All

Alexis A. Aronowitz

Facts: On 25 September 2021, seventeen-year-old William Chambers was convicted of the armed robbery and aggravated battery of Jackson Acer in front of Allen Public High School. The facts are undisputed. On 23 January, Mr Acer was walking home from work at approximately 20:25 when he was confronted by the defendant, Mr Chambers. Mr Chambers pulled a knife on Mr Acer and demanded that Mr Acer hand over his watch and wallet. When Mr Acer refused to hand over his watch, an altercation ensued during which Mr Acer was stabbed twice in the chest. Mr Chambers fled the scene of the crime with Mr Acer's wallet. The incident was captured on closed-circuit cameras and Mr Chambers was arrested two days later. After having been tried and convicted, all parties have returned to court for the sentencing hearing. The sentencing hearing is scheduled to determine to what degree William Chambers, as a minor, is responsible for the crime and the appropriate sentence. What transpires next, are the arguments presented by both the State (Ms Muirhead represents the Prosecution Department) and the defendant's lawyer (Ms McCormick), as well as testimony of the victim, Mr Acer.

Judge Falcone: 'Mr Chambers, you were found guilty by a jury of your peers earlier this month of armed robbery and aggravated battery resulting in serious bodily injury. We have now arrived at the sentencing hearing. I will explain to you what will happen next. Both the prosecution department and your defence attorney will present evidence to try and explain the circumstances of the offense and your background in order to help this court in reaching the most appropriate sentence for you. It is the practice of this court to hear victim impact statements. This means that Mr Acer will be asked to explain to the court what impact this crime has had on his life. I will also consider a report from the Probation service in helping me reach a sentence. After all of this, you will be allowed to make a final statement. Mr Chambers, do you understand what will transpire over the next two days?'

William Chambers: 'Yes, your honour, I do.'

Judge Falcone: 'In that case, as the representative of the Prosecution Department, Ms Muirhead, you may proceed.'

Ms Muirhead: 'Your honour, Mr Chambers' trial was waived to adult court because of the seriousness of his offense. Do not be swayed by the fact that the defendant is still a minor, or the fact that the defence team tries to evoke sympathy by calling him Billy. Mr Chambers is a dangerous offender, one who was armed, impulsive, and quick to use violence. While he may still be a minor, adolescents in this State, when involved in serious, violent crimes, can be charged and sentenced as adults.[1] If William Chambers did not want to be tried and sentenced as an adult, he should not have committed the crime.

'Your honour, the prosecution department has proven beyond a reasonable doubt that Mr Chambers robbed and stabbed Mr Acer on the night in question. Mr Chambers is a habitual offender. He has had run-ins with the police since the age of twelve. Up until this most recent crime, all of Mr Chambers' violations were property offenses. Regardless, Mr Chambers has shown a willingness to use violence and a persistent blatant disregard for the law.

'If one accepts the premise of rational choice, one believes that human beings exercise free will, and choose all behaviour, including criminal behaviour.[2] Behaviour involves weighing the criminal gains against the cost and risk of getting caught. Mr Chambers made a concrete decision to commit a robbery on the night of 23 January. We heard during the trial that William Chambers' motivation was not to injure Mr Acer, but as a consequence of the act of robbery, Mr Chambers must accept the consequences following the robbery – the injury of the victim.

'What is a suitable punishment for an offender who commits a crime resulting in the serious injury of the victim? Punishment should be severe, swift, certain, and proportional to the crime.[3] It should deter not only William Chambers from committing another crime but should also send a message to others in the community that crime will result in arrest, prosecution, and punishment. This is the principle of both specific and general deterrence[4] and underpins many sentences handed down by courts in this country. Other rationales for punishment include incapacitation and rehabilitation. And let us not forget retribution. An eye for an eye, a tooth for a tooth. Mr Chambers *deserves* to be punished for what he did to Mr Acer. All of these goals can be served by incarcerating Mr Chambers. He deserves to be sent to prison for the maximum sentence specified in this State's sentencing guidelines.'

Judge Falcone: 'Ms Muirhead, if the Prosecution department has nothing else to add, the court would now like to hear from the defence team.'

Ms McCormick: 'Thank you, your honour. The defence team disagrees with the Prosecution's argument that criminal behaviour is a rational choice. There are factors beyond the control of the offender that contributed to his developing into the young man that is standing before the court today.

Much empirical research from the fields of biology, genetics, sociology, economics, and psychology, among others, have pointed to correlations between individual, psychological, family, and environmental risk factors that predispose a person to become involved in crime, in general, and violence, in particular.[5] I would like to argue for leniency in the case of my client, Billy Chambers, by explaining to the court what factors in Billy's background contributed to his development into the young man he is today.

'The study of foetal origins has taught us that the first nine months of life, in the mother's womb, are critical to the development of the brain.[6] Stress and environmental contaminants, such as alcohol, drugs, or chemicals from cigarette smoke have been linked to difficult temperament and impaired cognitive functioning, both risk factors for delinquent behaviour.[7] During her pregnancy, Billy's mother smoked, consumed drugs, and abused alcohol. Billy, born premature, was rejected by his mother and the important bonding between mother and child was strained due to Mrs Chambers' postpartum depression.[8] Billy's mother continued to suffer from depression. Billy's father abandoned the family shortly after Billy was born. Due to her mental disorder, Billy's mother was incapable of taking care of him and throughout the first three years of his life, Billy was under the care of Child Protective Services and placed in an institution. This failure to bond with a caring mother has been linked to anxious or insecure attachment in infants.[9] Poor attachment to parents[10] is considered to be a risk factor for later involvement in delinquent behaviour.[11]

'Billy's mother took him back into her house when Billy was three. He was raised in a household deprived of books, educational toys, and a stable role model. His mother alternated between being a harsh and inconsistent disciplinarian, and neglectful parent.[12] She lived with an assortment of different men, some of whom were abusive towards her and Billy. At the age of seven, Child Protective Services petitioned Family Court to have Billy removed from her home and put in foster care. Billy reported to Child Protective Services, during monthly visits, that his mother's partners sometimes beat her and threatened him. He claims to have been sexually abused by two of his mother's partners. Billy moved three times before he was allowed to return permanently to his mother at the age of nine.

'Your honour, a study by some of the leading experts in the development of juvenile delinquency shows that "individuals who had been physically abused in the first five years of life were at greater risk for being arrested as juveniles for violent, nonviolent, and status offenses".[13] Research has shown that children growing up in households with spousal violence experience impaired cognitive and emotional functioning, demonstrating depression and aggression[14] and many researchers conclude that "children's exposure to domestic violence may generate attitudes justifying their own use of violence to solve problems and deal with frustrations".[15] The problem is exacerbated here by Billy's polyvictimization – "a term used to describe exposure to multiple types of victimization."[16] In a study examining the effects of polyvictimization, researchers found polyvictimization was strongly related to both emotional down-regulation (i.e., numbing and callous unemotional traits) and emotional dysregulation (i.e., dissociation and borderline traits) – and through them, to more severe delinquency – equally for boys and girls.[17]

'It is clear from Billy's exposure to violence as a child, and his mother's erratic and neglectful parenting, that Billy was subjected to multiple forms of child maltreatment and neglect. A large-scale study by Dong and colleagues found that multiple types of child maltreatment[18] often co-occur rather than occurring independently and that they accumulate and result in more adverse outcomes and a more severe psychopathology than if these types of maltreatment were isolated.[19] Other scholars found that "the underlying mechanism of the cumulative classification approach is that individuals with more childhood adversity may have biological and cognitive changes that may contribute to a lower response threshold to future stressors. Thus, this lower threshold makes these individuals more reactive to adverse experiences."[20]

'Billy's problems continued. Both Billy's mother and the schools he attended report that Billy was a difficult child. He often exhibited temper tantrums and was aggressive towards other children. While never having been diagnosed with learning problems or disabilities, his foster parents and school teachers report that he had problems concentrating on work at school and was often fidgety and hyperactive. Billy did poorly in school and when he was older he was often truant. He reported being bored and disliking school. He failed two grades and at the age of sixteen, stopped attending school altogether.

'Billy reports being rejected by most of his peers throughout his school time. At the age of fourteen, and prior to having left school, he reports hanging around the house and playing video games. He met a number of older boys at the corner bar and began associating with them. It was then, and through them, that he became involved with drugs. Your honour, Billy claims it was his drug addiction that led him to commit the robbery.

'On behalf of my client, I ask the court for leniency. A leading expert, Kenneth Dodge, introduced the Dynamic Cascade Model[21] of the development of violent behaviour. This model, based upon empirical research, identifies a trajectory into violent behaviour starting with an "adverse context". This was firmly established when Billy was born with cognitive impairments as a result of his mother's drug and alcohol consumption during her pregnancy and the failure of Billy to establish a warm, nurturing relationship with his mother in the first two years of his life. Her adverse parenting style and Billy's exposure to violence in the home further contributed to his slide into delinquency. This was further exacerbated by his poor school performance, conduct problems, and school failure. His absent mother neglected to monitor his behaviour and friends, and Billy fell under the influence of delinquent peers. Your honour, Billy Chambers is a textbook example of a young man who was destined to become involved in violent behaviour. Due to so many factors beyond his control, particularly during the early years of his life, I ask you to be lenient in sentencing my client.'

Judge Falcone: 'Ms McCormick, if you are finished addressing the court, I would like to give Mr Acer the opportunity to tell the court how he has experienced the offense.'

Mr Acer: 'Thank you your honour. I would like to direct my comments to Mr Chambers. Mr Chambers, you have no idea what your thoughtless act has done to my life. I hope that by sharing this with you, you will understand the serious nature of your assault and think twice before you consider robbing anyone else.

'On the night that you robbed me, I had just come back from church. You see, it was the anniversary of my father's death. The watch you tried to steal from me was his. It was never about the financial value of what you wanted to take from me, but the sentimental value. That watch could never be replaced. I gave you my wallet and tried to explain to you why I did not want to give you the watch. But you wouldn't listen.

'I spent nine days in the hospital. This was a tremendous burden on my wife and children. My wounds have healed and the physical pain has disappeared. But I have been diagnosed with PTSD. I have nightmares and often wake up in the middle of the night with heart palpitations and sweating. I have had problems going back to work and functioning normally. I've been unable to work full-time since the assault, which has had a dramatic impact on our family income. I can no longer work the night shift. I'm afraid to go out when it's dark and am fearful of wearing the watch I so love and which reminds me so much of my father. You have stolen from me more than a wallet and some cash. You stole from me a sense of being safe. You have no idea the impact that your selfish, thoughtless act has had on my life.

'Your honour, I have nothing left to say.'

Judge Falcone: 'Mr Chambers, before I allow you to address the court, I do want to review with you the presentence report drawn up by the Probation Department. According to the presentence report, you appear to have difficulty submitting to authority and adhering to societal expectations. Your previous behaviour has brought you into contact with numerous social and law enforcement agencies and you have consistently been involved in antisocial behaviour since you were a young adolescent. What does speak in your favour is that you have only been involved in property offenses, until this last crime.

'The presentence report details drug and alcohol abuse, which you claim to be the root cause of your problematic behaviour. I see also, that you have expressed deep remorse for the harm you caused to Mr Acer. You have been actively participating in group therapy while in pre-trial detention and seem to want to turn your life around. Is that a fair summary of your progress and development since your incarceration prior to trial, Mr Chambers?'

William Chambers: 'Yes, your honour.'

Judge Falcone: 'Mr Chambers. The last word is reserved for you. Do you have anything else you would like to add before the court is recessed?'

William Chambers: 'Thank you, your honour. I have not had an easy life. I would love to blame my mother for disadvantaging me even before birth, but I also have to take responsibility for the choices I have made. I've made many mistakes in my life and I now see the harm I caused Mr Acer. I am truly sorry, Mr Acer. I can never take back what I have done to you, but I can try to change and make something of my life, if the court will give me a chance. I need help with my addiction and managing my anger. I hope that I will be able to get this help.'

Judge Falcone: 'If you have nothing else to add, Mr Chambers, the court will now adjourn, and we will meet next Tuesday at which time you will hear your sentence.'

Epilogue

This article exposes students to the study of criminal behaviour through models and theories from the fields of criminology, sociology, psychology, biology, victimology, and law. The defendant is a male, because according to official statistics, almost 80 per cent of all violent crimes (in the U.S. in 2019) were committed by males.[22] Alcohol and drugs are often linked to violent behaviour.[23] We know, through the study of victimology, that the psychological impact on victims of violent crimes extends long after the physical injuries have healed.[24]

Studies from biology, in particular, the field of foetal origins, have shown that foetuses exposed to environmental contaminants in the womb, can suffer from cognitive impairments, a risk factor for later aggressive behaviour. The Cumulative Classification Approach holds that multiple forms of child maltreatment can accumulate and result in more severe psychological disorders and Dodge and colleagues argue that risk factors for juvenile delinquency lead to other risk factors for juvenile delinquency, the way a waterfall might cascade downhill. Biosocial theories[25] of crime tell us that the interaction between a biological predisposition and a negative environment is the strongest predictor of violence than either of these factors alone. I have created a character who has been exposed to multiple risk factors, both biological and social/environmental. Mr Chamber's attorney has used all of these risk factors as mitigating factors, to argue for a lighter sentence.

The prosecution department, having proven Mr Chamber's guilt beyond a reasonable doubt, introduced the rationales or principles of punishment, arguing for incarceration. As a habitual offender, studies have shown that chronic offenders are responsible for a large proportion of crimes. Birth cohort studies have shown that the chronic offenders (6% of the population studied) can be responsible for as much as 50% of all crime perpetrated by that population.[26]

After hearing all of the arguments, how would YOU sentence Mr Chambers?

Suggestions for further reading

J.C. Barnes, A. Raine, D.P. Farrington. 'The Interaction of Biopsychological and Socio-environmental Influences on Criminological Outcomes'. *Justice Quarterly* (2020), 1-26.

D.P. Farrington. 'Predictors, Causes, and Correlates of Male Youth Violence'. *Crime and Justice*, 24 (1998), 421-75.

A. Raine. *The Anatomy of Violence: The Biological Roots of Crime*. New York, NY: Pantheon, 2013.

Handling tricky questions

Jocelyn Ballantyne

Language is as integral to the human experience as air: it is all around us, something most people never think about deeply. Language is, at the same time, one of the human species' most powerful abilities. With a coordinated series of vocalizations, or characters on the screen of your computer or smartphone, you can put ideas into other people's minds. You can make others imagine an elephant, just by saying 'Imagine an elephant', and then make them imagine that elephant being ridden by, for example, Donald Trump, by adding 'Donald Trump is riding the elephant'.

Public figures of all kinds, including divisive politicians like former US president Donald Trump, make use of language's power to serve their strategic aims. They can convey their ideas not only with statements, but also with questions. Leading and loaded questions are tricky questions that advance the agenda of the questioner and are often used, in debates, in interviews, and in questionnaires, to provoke or manipulate. The linguistic mechanisms behind tricky questions are not specific to English, but are arguably universal, and operate in questions in every human language, from French to Chinese to Quechua. Just as biologists use theory and tools to understand how a particular kind of virus works, we can use linguistic theory and tools to understand how these kinds of questions work.

The examples in 1 and 2 come from a list of questions that the Trump transition team provided to the office of the United States government responsible for international aid before President Trump took office in 2017.[1] If you are like most people, these questions put the Trump team's hostility to international aid for Africa into your mind. (Joseph Kony is the leader of the Lord's Resistance Army, a guerilla group in Uganda; the US withdrew support for the manhunt in April 2017, three months after President Trump took office. PEPFAR stands for President's Emergency Plan for AIDS Relief, a programme established by president George W. Bush in 2003. Al-Shabaab is an Al-Qaeda affiliated militant group operating in Somalia.)

1　(a)　We've been hunting Kony for years. Is it worth the effort?
　　(b)　Is PEPFAR becoming a massive, international entitlement programme?
　　(c)　Is PEPFAR worth the massive investment when there are so many security concerns in Africa?

2   (a) Why do we support massive benefits to corrupt regimes?
    (b) Why should we spend these funds on Africa when we are suffering
         here in the US?
    (c) We've been fighting al-Shabaab for a decade. Why haven't we won?
    (d) With so much corruption in Africa, how much of our funding is
         stolen?

These questions represent only a few examples of the thousands of leading
and loaded questions that were posed in documents, public speeches, and
tweets by President Trump and his close associates in the years he was in
office. A leading or loaded question is never 'just a question', although that
is probably exactly what Mr Trump would claim if someone challenged his
use of them. This paper does not address the content of the Trump team's
questions, something better left to political theorists and historians; rather,
it explains the linguistic mechanisms that allow tricky questions to put
thoughts into the minds of an audience. Understanding these can help you
identify leading and loaded questions, and come up with responses that
avoid the traps they set.

How questions work

The first matter to consider in looking at how questions work is their form,
what linguistics calls *syntax*. One universal property of questions is that
their form systematically relates to the form of a corresponding statement.[2]
This systematic correspondence exists in all languages, despite the different
forms, or syntax, questions might have.

Consider first the type of question that is often regarded as the simplest
type: the yes-no question. Linguists typically call these polar questions,
since their possible answers are at the opposite end of the metaphorical
pole *yes-no*. The questions in 1(a)-(c) above are examples. We can see how
polar questions relate to corresponding statements by comparing examples
from two very different languages, English and Mandarin. The Mandarin
examples appear in the Roman alphabet, rather than Chinese characters,
and are accompanied by a word-by-word gloss, so you can follow it even if
you do not know Chinese.

3   English
    (a) Question: Are you a college student?
    (b) Statement: You are a college student.

4 Mandarin

		Nǐ	shì	dàxuéshēng	ma
(a) Question:		you	be	college student	QPART
(b) Statement:		Nǐ	shì	dàxuéshēng	
		you	be	college student	

The form of the polar question in 3(a) relates to the form of the corresponding statement in 3(b) in a systematic and predictable way: the question contains the same words as the statement, in a different order. The verb *are* appears at the beginning of the question, before the subject *you*, instead of after it, as in the statement. In English, the fronting of an auxiliary verb is the syntactic marker of a polar question. Likewise, the form of the polar question in 4(a) relates to the corresponding statement in 4(b): the question contains the same words as the statement, followed by the question particle *ma*. In Mandarin, this particle in this position is the syntactic marker of a polar question. Consciously or not, speakers always link the form of a polar question to its corresponding statement; the question, in essence, puts the statement in the minds of listeners or readers.

The syntactic correspondence of question to statement also holds for more complicated types of questions, called information questions. The questions in 2(a)-(d) above are examples of information questions. Examples in English (5) and Mandarin (6) again show how information questions relate to their corresponding statements. In English, an information question is introduced by a prominent information question word, *what* in 5(a), along with the fronting of the auxiliary verb *are* that we already saw in 3(a). A Mandarin information question has the same word order as its corresponding statement, but contains an information question word, in 6(a) *shénme*. Information questions are more complicated than polar questions because they relate to corresponding statements that are missing something, represented in 5(b) and 6(b) with the placeholder *X*. In 5(a) and 6(a), the missing bit is the object of the verb 'study', in fact what the questions are asking about.

5 English
   (a) Question: What are you studying?
   (b) Statement: You are studying X.

6 Mandarin

		Nǐ	zài	xuéxí	shénme
(a) Question:		you	AUX	study	what

Recognizing that the form of a question puts the form of a corresponding statement into the mind of the audience hints at how questions work in terms of meaning, or *semantic* function. Question meaning is arguably even more universal than question form.[2] As the examples from English and Mandarin show, the form of statements and questions differ in different languages, in words and in word order. These examples also show, though, that two different forms can mean exactly the same thing. That is, 3(a) *You are a college student* and 4(b) *Nǐ shì dàxuéshēng* have the same meaning, as do all the other comparable pairs of expressions in examples 3 through 6 that are translations of each other; this is, of course, what makes them translations. This seemingly trivial observation reveals a profound fact: the same general principles and mechanisms underlie linguistic meaning in all languages, including the meaning of questions.

The meaning of a question relates to the meaning of a corresponding statement, just as the form of a question relates to the form of a statement. To see this, let's begin again with the simplest type of question, the polar question. Although the polar questions in 1(a)-(c), 3(a) and 4(a) ask different questions, they share the same semantic function: each inquires if the meaning of the statement it activates is true or not. The insight, then, is that the basic meaning of a polar question queries two possible versions of reality; that is, we ask the question *Are you a college student?* when we can imagine a reality in which the person we are talking to is a college student, as well as one in which they are not.

The semantics of information questions is more complicated, first of all because there are many kinds of information questions, containing words like *what, who, when, why, how* or their equivalents in other languages. Secondly, information questions seek information that actually fills in the meaning of the corresponding statement. Two tricky information questions are repeated here.

2   (a) <u>Why</u> do we support massive benefits to corrupt regimes?
    (d) With so much corruption in Africa, <u>how much</u> of our funding is
         stolen?

The words *why* (2(a)) and *how much* (2(d)) are syntactic markers of information questions. These also contribute to the semantics of the question: *why* signals a query about the reason that *we support massive benefits to corrupt regimes*, while *how much* signals a query about which amount *of our funding is stolen, with so much corruption in Africa*. Like polar questions, then, information questions activate the meaning of corresponding statements. But while a polar question inquires whether a corresponding statement is true, an information question actually requires that the statement be true in the context for the question.

What tricky questions do

To see how the workings of questions contribute to their becoming leading or loaded, we turn now to *pragmatics*, the interaction of context with linguistic meaning. Straightforward questions function to request information: polar questions check whether a statement is true or not, and information questions seek additional information about something already partially known. Basic ground rules apply for asking straightforward questions: we ask a question when (A) we want information we do not already have, and (B) we reasonably believe our audience can provide that information.[3] Tricky questions break these basic ground rules. They do not function as straightforward requests for information, but as provocations that serve the questioner's rhetorical and strategic purposes. They exploit the interaction of question meaning with context to create communication danger zones that can be difficult to navigate.

Leading by implicature

Leading questions break ground rule A for straightforward questions: they do not sincerely ask something that the questioner wants to know. The questions in 1(a)-(c), repeated here, clearly transmit what the Trump transition team considers to be the 'right' answers (no, yes and no, respectively).

1 (a) We've been hunting Kony for years. Is it worth the effort?
   (b) Is PEPFAR becoming a massive, international entitlement programme?
   (c) Is PEPFAR worth the massive investment when there are so many security concerns in Africa?

Leading questions transmit the questioners' desired answers via a phenomenon called implicature. An implicature is an idea that is not explicitly stated, but still strongly suggested.[4] An implicature of 1(a) is that *the search for Kony has gone on unsuccessfully for too long*, of 1(b) that *PEPFAR is not doing valuable work*, and of 1(c) that *PEPFAR costs too much money*. We can make these ideas explicit to see how naturally they fit into the context for the questions, as in 1'(a)-(c).

1' (a) We've been hunting Kony for years – too long! Is it worth the effort?
    (b) PEPFAR is not doing valuable work. Is it becoming a massive, international entitlement programme?
    (c) PEPFAR costs too much money. Is PEPFAR worth the massive investment when there are so many security concerns in Africa?

These additions make no difference to the questions, and certainly do not change the desired answers, demonstrating how strongly the ideas are implied. Still, they are genuinely only implied in the original questions. If the context is altered to explicitly exclude these implicatures, as in 1'(a)-(c), it changes the interpretation of the questions, and crucially, the desired answers.

1" (a) We've been hunting Kony for years, and it has succeeded in seriously reducing his operations. Is it worth the effort?
    (b) PEPFAR has saved millions of lives, and can save millions more. Is it becoming a massive, international entitlement programme?
    (c) PEPFAR has saved millions of lives, and can save millions more. Is it worth the massive investment when there are so many security concerns in Africa?

Where do the implicatures come from? They arise in part because of the connotations of words appearing in the question and preceding context: here, *for years* (1(a)), *massive* (1(b) and 1(c)), *entitlement* (1(b)), and *security concerns* (1(c)). The implicatures make it obvious that the members of the Trump transition team are not asking these questions out of a genuine interest in new information. They are thus not straightforward questions, but have another purpose: to challenge, rather than inquire about, the perspective of officials of the US government concerned with aid to Africa.

Loaded with presuppositions

Loaded questions break ground rule B for asking questions: the questioners do not reasonably believe that their audience can answer them directly. What allows information questions to become loaded is the fact that they include the meaning of corresponding statements. Here again are the information questions posed by the Trump transition team, followed (in italics) by their corresponding statements.

2" (a) Why do we support massive benefits to corrupt regimes?
We support massive benefits to corrupt regimes.

(b) Why should we spend these funds on Africa when we are suffering here in the US?
We spend these funds on Africa when we are suffering here in the US.

(c) We've been fighting al-Shabaab for a decade. Why haven't we won?
We haven't won our decade-long fight with al-Shabaab.

(d) With so much corruption in Africa, how much of our funding is stolen?
With so much corruption in Africa, X of our funding is stolen.

The statements reflect the assumptions that the Trump transition team makes in asking each question, assumptions containing other assumptions. Assumptions that must be true in order for the statements that contain them to be true are called presuppositions.[5] Unlike implicatures, presuppositions cannot be cancelled by additional context (see 1" above).

To illustrate, the statements in 10 are presuppositions of the *why* question in 2(a).

10 We support massive benefits to corrupt regimes.
We support benefits to regimes.
The benefits are massive.
The regimes are corrupt.

Such presuppositions 'load' questions so that it is no longer possible to answer them in a straightforward fashion. Question 2(a) asks for a reason, but if someone were to respond with a reason, for example 'Because it keeps the region stable', this would automatically accept the truth of all the presuppositions listed in 10. Points for the Trump transition team! On the other hand, if someone were to respond by explicitly denying the presuppositions of the question, ('We do not support massive benefits to corrupt regimes!'), they would appear to be defensively avoiding answering the question – which also wins points for the Trump transition team. Similar traps are set in the other information questions 2(b)-(d).

Leading versus loading

Often, leading questions are polar questions, and loaded questions are information questions, but polar questions can also be loaded with presuppositions.

The obvious example from the Trump transition team's list is 1(c), containing the presupposition that *there are so many security concerns in Africa.*

1   (c)   Is PEPFAR worth the massive investment when there are so many security concerns in Africa?

Again, whatever the answer, this presupposition survives, and denying the presupposition explicitly ('There are not that many security concerns in Africa!') puts the respondents in defensive conflict with their questioners. This presupposition also strengthens the implicature that the circumstances that the question asks about are problematic, further driving the leading nature of the question towards the desired answer 'No, PEPFAR is not worth the investment'. Question 1(c) is thus both leading and loaded.

Similarly, information questions can become leading questions. *Why* questions are particularly susceptible to this: 2(a)-(c) all anticipate the response 'No good reason'. Like the questions in 1, the *why* questions in 2(a)-(c) also contain implicatures that the circumstances they are asking about are problematic, making them leading as well as loaded. The context and question in 2(c) illustrate this nicely. *Why haven't we won?* presupposes that *we haven't won*, and the context creates an implicature that not having won is problematic. In 2"(c), this implicature is made explicit to show how naturally it fits (compare with 1") above).

2"   (c)   We've been fighting al-Shabaab for a decade, and we should have won by now. Why haven't we won?

Questions 2(a)-(c) are thus leading as well as loaded. Implicatures make questions leading, and the presuppositions of the statements activated by questions make them loaded.

Deflect and disarm

Even experienced public figures can be so flummoxed by leading and loaded questions that their responses play right into the agenda of the questioner. Yet there are strategies for responding to tricky questions that avoid the traps that they lay, and can even turn them further to your own ends. To deflect a leading question, you can explicitly deny its leading implicature, rather than giving a simple answer, as in the examples of responses in 11. These responses use implicature to suggest an answer, but prevent the issue from being reduced to a simple yes or no.

11 (a) We've been hunting Kony for years. Is it worth the effort?
Implicature: The hunt has been unsuccessful for too long
RESPONSE: The hunt for Kony has succeeded in seriously reducing his operations.

    (b) Is PEPFAR becoming a massive, international entitlement programme?
Implicature: PEPFAR does not do valuable work
RESPONSE: PEPFAR has saved millions of lives, and can save millions more, for a few dollars per life.

    (c) Is PEPFAR worth the massive investment when there are so many security concerns in Africa?
Implicature: PEPFAR costs too much money
RESPONSE: PEPFAR has saved millions of lives, and can save millions more, for a few dollars per life.

To disarm a loaded question, respond with a statement that undermines the presuppositions that you do not accept, without explicit denial. You can effectively use implicatures to undermine the presuppositions of loaded questions. The hypothetical responses in 12 counter the presuppositions, without going on the defensive.

12 (a) Why do we support massive benefits to corrupt regimes?
Presupposition: We support massive benefits to corrupt regimes
RESPONSE: We provide support to governments whose policies make them a source of stability in the region.

    (b) Why should we spend these funds on Africa when we are suffering here in the US?
Presupposition: We spend these funds on Africa when we are suffering here in the US
RESPONSE: Humanitarian aid to Africa stabilizes key regions with resources important for US industries, and by extension, the US workforce.

    (c) We've been fighting al-Shabaab for a decade. Why haven't we won?
Presupposition: We haven't won our decade-long fight with al-Shabaab
RESPONSE: Military intervention has been keeping al-Shabaab in check, despite its being the largest jihadist organization in the world.

    (d) With so much corruption in Africa, how much of our funding is stolen?
Presupposition: With so much corruption in Africa, X of our funding is stolen

RESPONSE: Financial support goes to organizations who meet international good governance transparency requirements.

A bonus benefit of using implicature to undermine presuppositions is that
you can follow up with explicit denial of problematic presuppositions without
sounding defensive. For example, a US official could continue the response
in 12(a) with: 'We do not support massive benefits to corrupt regimes.' Try
this yourself for the other examples in 12. Rhetorical points for you!

Whether the aim is to share information or to further a strategic agenda,
the power of language comes from its ability to put ideas into the minds
of others. Unpacking the linguistic mechanisms that contribute to tricky
questions gives insights relevant for constructing responses that avoid
their rhetorical traps. As we have seen, questions of all kinds correspond to
statements, and activate the meaning of those statements. Leading questions
exploit implicatures generated by the context to push towards a particular
answer, while loaded questions exploit the presuppositions of the statements
they activate. Important to deflecting a leading question is thus identifying the implicatures involved, so that these can be explicitly cancelled.
Disarming a loaded question requires identifying its presuppositions and
undermining them, with implicature, in order to avoid explicit denials that
make a response seem defensive.

Language is a powerful human tool. As with many tools, it is possible to
use it, and even to use it well, without understanding what makes it work.
Understanding the mechanisms of this powerful tool, though, gives us an
advantage in fending off ideological adversaries and protecting ourselves
from becoming pawns in their strategic aims.

Suggestions for further reading

H.P. Grice. 'Presupposition and Conversational Implicature'. Ed. P. Cole, *Radical
    Pragmatics*. New York, NY: Academic Press, (1981), 183-198.
L. Karttunen. 'Syntax and Semantics of Questions'. *Linguistics and Philosophy*,
    1(1) (1977), 3-44.
R. Stalnaker. 'Pragmatics'. Eds. D. Davidson, G. Harman, *Semantics of Natural
    Language*. Dordrecht: Reidel, (1972), 389-408.

What is Meaning?

Gaetano Fiorin

At this very moment, your eyes are following a sequence of black symbols imprinted on a white page (or, alternatively, a luminescent screen) and transmitting the image they capture to the brain. In turn, the brain activates a complex network of neural circuits and elaborates the symbols into a meaning – information that is significant to you as a cognitive agent. The question I will address in the coming pages is this: what is this thing that is reaching you at this very moment through the mediation of a sequence of symbols? That is, what is meaning?

No doubt, this is a difficult question. Meaning appears to be a rather elusive and abstract notion. Yet, it is as much apparent that the ability to convey meaning through language is one of the most fundamental features of our shared human nature. Meaning is everywhere, an endless flow of information that accompanies us throughout our lives. Its role is as central to the privacy of our minds as it is to the public arena of our social relations, part and parcel of our daily activities and routines, including that of reading this page, right here, right now. It seems only fair, for an inquisitive mind, to aspire to know what such an essential feature of who we are and what we do is all about.

Throughout history, thinkers of different schools have approached meaning in different ways, investigating its many dimensions and delivering different answers. In what follows, I will briefly illustrate two of the most common views on the nature of meaning and will rely on them to draw some general conclusions about the cognitive system we call language.

Meaning as reference

The first view is one that has a long history in Western philosophical thought, at least since Plato. See especially Plato's *Theaetetus*. The most explicit formulation of the thesis of meaning as reference in the Western tradition is by the early Christian philosopher Augustine of Hippo (see in particular book I of his *Confessions*) and for this reason is sometimes referred to as the Augustinian view of language (a notable example is Wittgenstein and Anscombe's *Philosophical Investigations*.[1] It is commonly known as the thesis of meaning as reference and is based on the following simple claim: the meaning of a linguistic expression is the object it refers to. As an example, consider the name 'Plato'. This name belongs to the linguistic repertoire of most speakers of English both as a sequence of sounds, when pronounced, and as a sequence of letters, when written. What is its meaning? According to the thesis of meaning as reference, its meaning is the object it refers to, that is, the Greek philosopher who was born in Athens around 428 BC and was a pupil of Socrates and teacher of Aristotle.

The most appealing aspect of this thesis is that it draws a straightforward distinction between meaning and the language that expresses it. According to the thesis, meaningful language consists of two distinct domains of objects: on the one hand, the expressions of the language, things such as the proper name 'Plato', understood either as a sequence of sounds or a string of written symbols; on the other, the references of these expressions, the objects linguistics expressions are about – things such as the Greek philosopher Plato. Reference is the glue that holds these two worlds together.

It is in virtue of such distinction that the thesis allows us to identify a central property of meaning – the fact that it transcends the concrete means through which it is expressed. Meaning is conveyed through language but is of an altogether different fabric. The language we use to convey meaning is one thing, made of concrete events such as sounds or written symbols. The meaning these material means convey is just something else. Without doubt, this is one of the most fascinating features of meaning: whereas the very existence of meaning depends on the material means that convey it, its significance reaches far beyond them.

In its long history, the thesis of meaning as reference has met many objections. Two of them are worth mentioning here.

To begin with, we should acknowledge that the thesis of meaning as reference is but a modest first step in the long path towards answering the question of what meaning is. The thesis, as we said, claims that meaning is what language is about. This, however, raises another question: if meaning is what language is about, what is language about? This, it turns out, is an even more difficult question than the one we began with. It is, indeed, an issue over which thinkers of all walks of life have confronted each other for as long as we know.

One of the main points of dispute concerns whether the things language is about are concrete objects belonging to the natural world or thoughts and ideas belonging to the realm of the mind. Plato himself was convinced that the words we utter must ultimately refer to a reality that exists independently of our minds. In contrast, Aristotle believed that words are symbols of 'mental experiences'. Since then, these contrasts have characterized much philosophical reflection on language and are still very much alive today. In the twentieth century, when philosophers such as Willard Van Orman Quine[2] and Hilary Putnam[3] were elaborating philosophical arguments that meaning cannot be a psychological object but a dimension of social behaviour, cognitive scientists such as Noam Chomsky[4] were demonstrating that our understanding of the world is always and necessarily filtered by the complex and largely unconscious apparatus of our cognition.

The question remains far from settled. On the one hand, our everyday use of language suggests that language is as suitable to talk about the world around us as it is to talk about our inner psychological lives. On the other hand, if meaning is a purely psychological object, it remains a mystery how we manage to share it in the public arena of our conversations and use it to make provable factual statements. Obviously, the issue has profound philosophical implications, concerning our understanding of the world around us, our minds, the interplay between them, and the role of language in such interplay.

The second shortcoming of the thesis of meaning as reference I wish to mention is of a more technical linguistic nature. The thesis of meaning as reference works especially well with expressions such as proper names. For example, the reference of the name 'Plato' can be easily identified with the famous Athenian philosopher. Things, however, become more complicated as soon as we try to extend the same idea to other grammatical categories. Consider, as an example, the verb 'walk'. Surely it carries a meaning as much as the name 'Plato' does, but what is this meaning exactly? The thesis we are considering encourages us to think about the meaning of 'walk' in terms of the object it refers to, but what could possibly be the object referred to by a verb? Things become even more problematic with words such as the preposition 'of', the article 'the', the quantifier 'every', the relative pronoun 'who', the coordinating particle 'and', the subordinating particle 'if', and the negation 'not'? What could possibly be the reference of these words? As linguists know very well today, the meaningfulness of words such as prepositions, articles, quantifiers, relative pronouns, coordinators, subordinators, negations, and other similar categories has nothing to do with what they refer to. It has rather to do with the function they perform within the grammatical structures in which they occur.

Taken together, these observations underscore a more general problem. Words are not just individual units – separated islands each providing their own individual meanings. A significant part of the competence speakers have of their language consists in the capacity to combine words together in larger grammatical structures to convey meanings of higher and higher complexity. As modern linguistics teaches us, language is a creative machinery, which allows us to use a finite pool of resources to produce a potentially infinite number of meanings.

Meaning as an emergent property

The observation that language is a combinatorial system is a central premise to the second view of meaning I wish to illustrate. This view finds its roots in the work of the linguist Ferdinand de Saussure and claims that meaning is an emergent property of the structure of the language. To appreciate this definition we must first understand the terms 'emergent property' and 'structure'. I shall begin with the latter.

At the beginning of the twentieth century, Saussure initiated one of the most fundamental paradigm shifts in linguistics by contemplating the notion that language is a system of grammatical relations.[5] Words are not isolated units but nodes in the larger network we call language. According to Saussure, in fact, the system of a language can be captured in its entirety as a vast network of nodes and connections: the nodes are the words of the language; the connections are the grammatical relations that hold between them. To refer to this network, Saussure used the term structure and, for this reason, his theory came to be known as Structuralism.

In a structuralist framework, meaning is understood as an emergent property of linguistic structure. The term 'emergent property' is borrowed from the natural sciences where it is typically used to refer to the properties of a physical system that characterize the system as a whole but not its individual constituents.[6] An example of emergent property is the state of a system of particles. A body of H_2O molecules, for instance, can be in three different states: solid (in which case we call it 'ice'), liquid ('water'), or gaseous ('vapour'). What determines the state of the body is how rigidly the molecules in the body are tied to each other. If the bounds between the molecules are very rigid, we have ice; if they are looser, water; if they are even looser, vapour. We say that the state of the body is an emergent property because it has nothing to do with the molecules themselves. No matter the state of the overall body, the molecules and their elements are always the same – Hydrogen and Oxygen. What determines whether the body is ice, water, or vapour is the way these molecules are held together.

Within the structuralist paradigm initiated by Saussure, meaning is regarded as an emergent property of the grammatical structure of the language. Words do not bear a meaning of their own. Rather, their meaning emerges from the way they interact with each other within the larger system of the language.

There is a useful parallel we can rely upon to better understand this view: music. Consider this simple exercise. Its goal is to produce a musical melody starting from the following simple instructions: (i) pick a note, any note, and repeat it three times at regular pace; (ii) move four half-tones below the first note (on a piano this would mean four keys to the left) and play that note once. Let us say we pick note C in step (i). The melody we produce from following the instructions is C, C, C, Ab. If you were to play these four notes one after the other at a regular pace, you would immediately recognize them as the celebrated incipit of Beethoven's Fifth Symphony. What is interesting about these instructions is that they work irrespectively of the note we pick in step (i). If we pick note G, instead of C, the same procedure delivers the sequence G, G, G, Eb which, if played, would be perceived as the exact same melody. The exercise demonstrates that what characterizes a melody, what we remember about it and whistle under the shower, is not the exact notes of which it is composed but, rather, the relations of time and pitch that hold between them. The significance of a melody is, henceforth, an emergent property of the pitch and time relations between the notes it comprises. Notes do not have such significance on their own, they only acquire it when put in relation to each other.

The structuralist framework invites us to think of the meaning of language in the same way. Meaning does not belong to words in the same way as a melody does not belong to its notes. It is only when we look at words as part of the grander system of grammatical relations that constitute a language that meaning emerges as an overarching property.

Structuralism has represented a fundamental breakthrough in the study of language. We owe to Saussure the fundamental realization that language and meaning cannot be regarded as two separated entities. The grammatical mechanisms at the core of the language machinery are pivotal not only to its formal organization but also to the construction of meaning.

However, despite these major achievements, the structuralist view of meaning view also has its shortcomings. In particular, it is clear that language is not really like music or, at least, not entirely so. Grammatical relations alone are not all there is to meaning, as demonstrated by the fact that knowing the properties of a complex grammatical structure is never a sufficient condition for also knowing its meaning, or, at least, not entirely. We can appreciate this point by considering the following famous example by the linguists Charles Kay Ogden and Ivor Armstrong Richards:[7]

The gostak distims the doshes.

This sentence offers a curious linguistic puzzle. To most English speakers, the sentence appears unintelligible, although, and this is the curious part, not entirely. We understand that the sentence describes a situation whereby something called 'gostak' does something called 'distimming' to some other things called 'doshes'. We are able to understand at least part of the meaning of the sentence because we are able to recover the grammatical relations between the words in it. We understand that 'gostak' and 'dosh' are nouns and that 'distims' is a verb. We also understand that 'the gostak' is the subject of the sentence – the one doing the 'distimming' – and 'the dosh' is the object – the one undergoing the 'distimming'. The meaning we attribute to this sentence is, henceforth, a function of the structure of the sentence. This suggests that Saussure was at least partially right. To know the meaning of a linguistic expression we must know its grammatical organization.

There is, however, a part of the meaning of the sentence that we cannot recover from its grammatical relations alone. It is not enough to know that 'the gostak' is the subject of the sentence, that 'the doshes' is the object, and that 'distimming' is the transitive verb that connects them. To fully comprehend the sentence, we must also know what sort of things 'gostaks' and 'doshes' are and what sort of action 'distimming' is. This is something that no speaker of English can know, because the words 'gostaks', 'doshes', and 'distimming' are, in fact, invented. They look like they could be English words but they are not. They simply do not belong to the vocabulary of any English speaker. Crucially, the meaning of these words is not something we can systematically recover from the structural relations between the words in the sentence. It is something that must be stipulated outside the realm of grammar and, ultimately, language.

Conclusions: Meaning as interface

There are a few conclusions we can draw from the previous discussion, some negative, others positive. The most important negative conclusion we seem to have reached is that none of the frameworks reviewed so far are truly satisfactory. In different ways, all of them appear to focus on one aspect of the notion of meaning while disregarding others that are, after all, as important. In particular, the view of meaning as reference focuses on the aspects of meaning that transcend language but completely disregards the fundamental link between meaning and grammar. The structuralist perspective, conversely, focuses exclusively on the role of grammatical structure in the construction of meaning but fails to capture the aspects of meaning that are independent of the language that conveys it.

Taken together, these negative conclusions pave the way to a constructive insight. Whatever meaning may be, it must be a multidimensional phenomenon characterized at once by the ability to refer to an independent reality and by the ability to organize such reality around the primitive mechanisms of grammatical structure. As a way to reconcile these two dimensions, I have proposed, in recent work with my colleague Denis Delfitto,[8] that language is an interface, a notion borrowed from the cognitive sciences and, in particular, the study of perception.

Perception, the way we see the world through our senses, has puzzled philosophers and scientists as much and for as long as that of meaning. Throughout the centuries, many have observed that the way we perceive things through our senses differs significantly from the way things really are and, on the basis of that, have questioned whether we should in fact trust perception as a reliable source of information (the philosopher René Descartes was a champion at formulating sceptical arguments of this sort). On the face of it, the work done by neurobiologists and cognitive scientists starting from the second half of the twentieth century has delivered a much more reasonable and less puzzling view of perception. According to these studies, perception is a system of interface between a perceiver and its environment. On the one hand, the ultimate goal of perception is that of capturing information from the perceiver's environment for the sake of delivering it to the perceiver. On the other hand, in order to achieve this goal, perception structures the information from the environment around categories that are chosen not only on the basis of how well they represent the environment but also of how compatible they are with the cognitive organization of the perceiver and its goals. In a way, perception strives to achieve a working balance between detecting environmental information in the most faithful way possible and packaging this information in the way that is the most intelligible and, therefore, useful to the perceiver.

We can think of language in the same way. The primary role of language is, as the view of meaning as reference contends, that of referring to a reality that is independent from language. However, in order to achieve this goal, language packages such reality in the form of categories and relations that are compatible with the user's cognition and, henceforth, of real epistemic value to the user. If this view is correct, language is but a cognitive function of a kind we already know: a system for translating a complex reality into intelligible information. Meaning is the product of such a function.

Suggestions for further reading

P. Portner. *What is Meaning? Fundamentals of Formal Semantics.* Oxford: Blackwell, 2005.

M. Morris. *An Introduction to the Philosophy of Language.* Cambridge: Cambridge University Press, 2006.

G. Fiorin, D. Delfitto. *Beyond Meaning: A Journey across Language, Perception and Experience.* Dordrecht: Springer, 2020.

II

Dialogue

Parmenides and Dōgen – an encounter

Chiara Robbiano

What follows is a dialogue, or multilogue, between two philosophers (the ancient Greek Parmenides, fifth century BCE, and the medieval Japanese Dōgen, thirteenth century CE), and a hiker who barges into their conversations, describes them as 'nondualists', and provides them with an opportunity to test their philosophical outlook against a concrete situation.

Nonduality refers to views – in philosophy, anthropology, linguistics, cognitive science, and other disciplines – that deny the ultimate reality of any fundamental duality, such as mind and body, you and I, my environment and I, subject and object. Nondualists argue that such dualities might sometimes be useful, but are ultimately conventional, human-made: they do not refer to separate entities.[1]

Some nondual philosophers might suggest focusing on the trustworthy fact of being conscious (or experiencing, or existing). They might argue that anything different from the fact of being conscious cannot be real, thus should not be feared. Parmenides uses this strategy,[2] as do Indian Advaita Vedānta philosophers, who defend a nondual interpretation of the Upaniṣad (last part of the Vedas).[3] Whereas Parmenides does not discuss the existential benefits of his nondual outlook, Advaita Vedāntins such as Śaṅkara, suggest dwelling in the trustworthy fact of one's self-consciousness, in the unshakeable conviction of one's stability and invulnerability, and behaving as an impassive watcher of whatever happens – since this attitude will bring liberation from suffering.[4]

I personally prefer *dynamic* nondual systems, such as Dōgen's Zen Buddhism, that deny that one can be a person on one's own. They assume that humans are not separate from the rest of reality but consist of relations with each other and with their environment (family, communities, geographical places, etc.).[5] Just like other nondualists, they also argue that any distinction might well be useful and necessary from some perspective, but points to a conventional boundary that can always be crossed.

This conception of nonduality allows for a fruitful dialogue. If we are intrinsically related to others, then real dialogues are possible. The participants in real dialogues welcome the possibility of being changed by their encounter with the other.[6] A nondual reality is a dynamic totality that is constantly recreated by the activity of each individual.

Philosophical dialogues may be found in very diverse philosophical traditions, for instance Greek, Indian, Chinese, and Japanese. They often feature people open to changing their minds. As we will see, Dōgen suggests that we are continuously co-creating our dynamic world, in dialogue with others, in response to what happens. Would you be open to listen to Dōgen, if you met him during a hike?

P: Hello Dōgen. Don't you think that the view from the top of this mountain is absolutely stunning?

D: Hello Parmenides. The glorious view, the crispy air and meeting you, my dear, make this moment uniquely pleasurable.

P: Look at the vast landscape in front of us: wherever you look, there are no gaps. Even if we have different words for lakes and forests, flowers and cherries, light and shadows – they all equally *are* (DKB9, B19, B8, 38-41).[7] It all *is*.

Dōgen spends a few minutes in silence, taking the vast landscape in; then says:

D: You can only have what 'fits' in the now. All we have fits in this node of time in which we are watching the landscape. Yes, we can say that all 'is'. We can even call it 'being'. But do you realize what has *dripped* into your being here right now? (*Shōbōgenzō, Uji*)[8] All the previous steps you have taken, all the mountains you have climbed have dripped into this now – which is all we have and what we are: learning how to walk, holding hands with your Dad, the shadows that soothed your eyes after walking in the blinding sun; the delights at bathing our feet in that cold lake last summer – do you remember? – which hurt like hell when we put them back into our hiking boots! So many experiences, memories, habits, and anticipations are here now together while we look at the landscape.

P: You can say it so beautifully and so evocatively. And yet I think we mean the same: 'it is now, all together, one, continuous' (DKB8, 5-6). It makes me feel so safe to know that 'all is full of *being*, thus [it] is all continuous' (DKB8, 24-25), this being that is knowing or experiencing (DKB3) is inviolable (DKB8, 48). Any experience, memory, anticipation, both lights and shadows: they are all being (B9), which we can access directly, immediately, trustworthily (DKB2, B3). This realization makes me feel so unshaken (*atremes*, DKB8, 4; B8, 26). You look so serene, Dōgen, I bet you have the same.

D: Well, we should explore this further. I might well agree that it is all here, continuous – but what you said about gapless 'experiencing' being invulnerable and making you feeling safe makes me suspect we might not be in complete agreement. You see: invulnerability smells of rigidity, lack of capacity for improvisation...

P: You know what – let's sit down under that tree and explore the issue further. A real dialogue with someone as clever and sensitive as you are is a rare treat.

A fellow hiker who has eavesdropped on their conversation suddenly barges into the conversation:

H: Gentlemen, forgive my intrusion. I am absolutely thrilled to have found my favourite nondualists on the top of my favourite mountain – I mean would you approve of me naming you as such?

They look at each other and then nod. Dōgen takes the floor and reassures her:

D: Yes, my lady, we both regard ourselves as part of the continuous experiencing of reality and not as subjects in front of objects, if that is what you mean.

She nods excitedly and continues:

H: Would you allow me to listen to your conversation for a while? Please...

Dōgen's kind eyes immediately open up in a benevolent smile. Parmenides resents the intrusion a little. He is fond of Dōgen to the point of being a little excited whenever he can spend some time with him. But he is a gentleman and a teacher at heart. They all go and sit together under the tree they spotted.

P: Dōgen, my dear, you mentioned a possible disagreement between us. But how could our nondual being – which is the same as experiencing (B3) – be less than invulnerable? How could it ever suffer any lack, or pain, if there is no second being, no 'other-than-being', no 'not-being' that could endanger it? You know my argument well (DKB2, B3) – any so-called not-being, any lack or absence, is just a name that we have given to some aspect of being. Of course, we need at least two principles – 0 and 1, night and day, man and woman – to found science, write poetry, to navigate our lives; but trustworthy reality is nondual, undivided, invulnerable.

The hiker's smartphone starts ringing and buzzing. She blushes, tries to put it on silent mode, but cannot help seeing the text message. She looks so dismayed that the two philosophers cannot neglect it. They politely ask her what the matter is.

H: It's my daughter. She lives on the other side of the world, you know. I miss her like crazy. She had planned long ago to come and visit me tomorrow. I was so excited that I decided to climb this mountain so that I would be very tired, and I could get some sleep tonight. I have bought two tickets for a concert tomorrow night. In my head, I kept playing the movie of being there with her, listening to Bach together, just like when she was tiny and we put on oboe sonatas, cello suites, violin concertos... Anyway... guess what?! She is going to bring her new boyfriend with her! That is so preposterous. I dislike men. They are so predictable. Creatures of habit. Cannot see anything

other than what they have ever seen or do what they have always done. *She blushes.* Of course, I am not talking about the two of you.
They spend a few long seconds in silence. Then Dōgen throws in an idea.
D: Parmenides, I have a proposition to make – the two of us will each analyse this situation and offer advice to our friend, from the framework of our respective philosophies. This might both help her deal with this sudden change of plans and help us single out the differences in our views.
P: Excellent idea, my friend!
Parmenides is a bit nervous. Perhaps because of his impatient Mediterranean temper, he takes the floor, supposing Dōgen will be fine with speaking second.
P: You see, my dear lady, there is neither a real past when your daughter lived with you nor a real future when you are going to listen to Bach, either the two of you or with an unexpected guest. There is no real daughter who is far away most of the time, or close by tomorrow. Do not cut off being from being with your mind (DKB4) – this will cause you anxiety, fear, false hopes; you will feel shaken by distinctions that can only be useful and innocuous if you are aware of their lack of reality. Experiencing is one and undivided; do appreciate this invulnerable certainty.
He says all this without catching a breath, without looking at either Dōgen's or the hiker's eyes. The woman and Dōgen give each other a puzzled look.
Then Parmenides gets out of his metaphysical frenzy, takes the woman's hand in his, with an almost fatherly gesture, looks at her pale complexion.
P: I think you have been neglecting yourself, my dear. Just like the moon – always looking at the rays of the sun (DKB15) – you have put your daughter at the centre of your universe. You need to learn to stand on your own two feet, with unwavering knees.
The hiker frowns.
D: Come on, Parmenides, you are embarrassing her. What do you mean?
P: The point is, that if you resent the workings of Eros, if you focus on boyfriends who come and go, you will not find peace. Look at it differently. Realize that your heartaches, your pain at missing her, your jealousy at the men who attract her attention – as much as your memories of holding her, your hopes for her future – are just unreal superimpositions on the only undivided being. The being you can trust right now is the same as your knowing (DKB3): your present immediate awareness, your experiencing in this very moment, no matter what angry words you would use to describe it. There is only one possible choice: the one for 'being' – there is nothing other than being. But you have to make this choice (B8, 15-16). If you don't, impotence in your breast will lead your scattered mind (B6) and shake you.

The hiker is on the verge of crying. Parmenides' words overwhelmed her. She can follow his reasoning, but she does not feel any better. His piece of advice seems to point at a heartless attitude. The resentment she feels at the change of plans announced by her daughter, who decided to bring her new lover along, is not something that can be considered as 'unreal' in order not to suffer. How can she bracket all things past, far away, and future as mere superimpositions on being, on the now?! She attempts a polite reply. But she is angry and disappointed.

H: Parmenides I appreciate your attempt, but this 'being' you want me to trust, this complete 'now' with neither a past nor future, rather than a place in which I can feel safe and invulnerable, feels like a place where I cannot breathe, for lack of air, and where I cannot love, for lack of others. When Nietzsche described you as *'blutlos wie eine Abstraktion'*[9] and as a spider who not only sucked blood but even hated the blood of its victims, since it smelled of experience, I got mad at him, and screamed that he did not understand that being is everything: it is the opening of flowers, parental worries, anticipation of summer, memories of lovers snuggling under a blanket – all this is being and all this is in the now. Today I wonder if Nietzsche was right after all. You are now telling me that flowers, parents, summers, and lovers are just superimpositions on invulnerable being...

Parmenides is taken aback. He tries to feel better by reasoning that he had never been at ease around girls – he always jokes with his friends calling them 'those born on the left side' (B17). Now he cannot help wishing she had never disturbed his day with Dōgen, to whom he now looks, waiting for his answer.

D: My dear, I appreciate your anger, disappointment, and confusion, your doubts. Parmenides suggests you single out what stays the same, what is invulnerable – and focus on trusting it, while disregarding the rest. But wouldn't you both agree that whatever we trust or focus on, we still need to *act*, in every precious, unique now?

The hiker nods hesitantly. Parmenides is puzzled, he had not seen this coming.

D: What I would encourage you to trust – or to revise – is not the naked fact of being, but your response to this situation and the values that inform it. How you act in the present moment is the blossoming of virtues that you have cultivated throughout your life – consciously or unconsciously. *Dōgen points at some blossoming flowers.* We are making this world at every moment. There is no far away which is separate from me now, there is no absence that I cannot carry with me in the now. You are right, Parmenides: the first step is becoming aware of being non-separate from the world around us. It is all being, trustworthily accessible by me right now, not endangered by any gaps of not-being. It's good to wake up to the continuity and the presence of all-exhaustive times and exhaustive worlds (*Shōbōgenzō, Uji*).

Parmenides regains a bit of colour on his cheeks.

D: But carrying the responsibility for world-making is what it is all about. The now is not complete in such a way that allows us to sit on our laurels. The now is open-ended, it is not yet, it is on its way towards something else which will result from the actions, the causes and conditions, the encounters that take place right now (*Shōbōgenzō, Uji*). *Now* you have encountered this text message that has changed the quality of your waiting for your daughter. *Now* you have encountered Parmenides' take on this message that has changed your waiting. *Now* you are opening up to my take on this text message, which will indubitably change the quality of your waiting. The time you will meet her will change your relationship with her (both in the past and the future), with Parmenides and me (which conventionally will have taken place in the past and might continue in the future), and your relationship with this very text message.

Both Parmenides and the hiker are now very attentively listening to him.

P: But, Dōgen, why should we give so much importance to all these fleeting moments, the emotions that we go through, and the silly human words we devise to cope with them? Are you suggesting we spend our lives in the dark, chasing fireflies?

D. Parmenides, all this might well be fleeting, but it is the stuff we and reality are made of. I could not agree more that our concepts and words – missing your daughter, liking Bach, longing for a mom-daughter moment – are devised by us humans, to cope with our little lives. However, rather than dismissing them, we should look for the values that hide behind them.

H: It seems very difficult to become aware of the values that inform our emotions and the concepts that we use to make sense of them. Why is that so important, Dōgen?

D: Because we shape reality with each of our gestures. The concepts we use are gestures, among other clumsy or graceful gestures, that we make and that keep shaping the open-ended world. Not only do we fragment reality along the lines of our concepts. More crucially, we shape it in a certain way depending on how we practise, how we move our body in tune or out of tune with reality, how we respond to any situation. The presence of an unexpected guest, your fear that he will claim all your daughter's attention, your immediate labelling all of this as a disappointment, an unwelcome change of plans, almost a betrayal, elicit a virtuosic response from you: you will need to improvise, perhaps to review your immediate reaction, in order to stay close to your values of openness and love, to respond wisely to the unique encounter of the three of you.

H: This makes sense. But, but... where to begin to be as wise as you are suggesting?

D: Let's see... in case of doubt, ask yourself, am I creating beauty or ugliness? Am I moving gracefully or clumsily? You create beauty and you move gracefully, when you express openness, attuned improvisation, by practising responding harmoniously and loving. Being continuous with reality means that we are responsible for the whole of it. We re-create it with each of our gestures. 'The opening flowers and falling leaves of the present are just the realization of continuous practice' (*Shōbōgenzō, Gyōji*). Moreover – come on! – not all guys are hopelessly stuck in their habits and beliefs.

They both glance at Parmenides, who does his best to be open to this expression of dynamic nonduality.

D: You know what? The new boyfriend might even be special. He will certainly change the dynamic of your encounter, but then again, every encounter is different and transforms the whole of reality every time. Thus, prepare to meet him not only by checking your mental attitude: the meeting must go through the body, which is the best candidate to express one's values. What about the three of you cooking together? See welcoming this guy as practice that is an attuned, responsive, and compassionate activity. This, I recommend to you rather than merely trusting 'being', which might well result in avoidance of encounters. Not unshakenness I recommend to you, but to be shaken gracefully.

Suggestions for further reading

J.W. Heisig, T.P. Kasulis, J.C. Maraldo, Eds. *Japanese Philosophy: A Sourcebook*. Honolulu: University of Hawaii Press, 2011.

D. Loy. *Nonduality: A Study in Comparative Philosophy*. Amherst, NY: Prometheus Books, 2012.

C. Robbiano. 'Can Words Carve a Jointless Reality? Parmenides and Sankara'. *Journal of World Philosophies*, 3(1) (2018), 31-43. https://scholarworks.iu.edu/iupjournals/index.php/jwp/article/view/1615

Ed. K. Tanahashi. *Moon in a Dewdrop: Writings of Zen Master Dogen*. London: Macmillan, 1985.

B.W. Van Norden. *Taking Back Philosophy: A Multicultural Manifesto*. New York, NY: Columbia University Press, 2017.

On being a doctor

Cornelus Sanders

'You need to unmute yourself', I blurt out as I watch my patient pantomime on the computer screen in front of me. Her face draws nearer as she scrutinizes her screen icons, juggling the microphone that slips off her head and when it hits the floor, a loud bang in my headset tells me that we have established an audio connection. It takes a moment of silence to recuperate and then we move on with one of the online video consultations that have replaced some of the regular patient visits to my office during the Covid-19 pandemic. The spatial boundaries between a remote doctor's office and the patient's private sphere have become fluid and blurred. In a literal sense, when some patients appear with fuzzy faces due to bad connections ('they told me the Wi-Fi had been fixed'), but also through multiple distractions, like jackhammers and sirens from the streets ('it is the big city, you know'), the pets and children parading across the screen ('so this is Janice and here is Ricky, no, not yet ready for school') and here is Henry, our cat, ('Oh don't walk over the keyboard, look what you have done'), as an error message pops up and ordinary household sagas ('What is that smell? Oops, sorry, have to run'), it pauses and interrupts our conversation. It is not only the stressors of maintaining our online communications, ordinary health issues, or disease threat that become apparent. Apart from the health concerns associated with my specialism, added tensions, at times oozing from the screen, are tangible as people are facing isolation and loneliness, balancing childcare and professional lives, losing jobs or even loved ones, and coping with uncertainty about the ways the pandemic will unfold. Many people show great resilience in coping with these kinds of challenges... but some are at risk of depression or post-traumatic stress.

Yet we seem to have slipped quite easily into this online patient-doctor interaction and apparently many patients are relatively pleased with the outcome so far. I find it striking that remote connections are proving satisfying because I sense there are several things amiss in the way we interact via two dimensional screens and unruly acoustics.

The basic premise of a good doctor-patient relation is trust, and this has to be created with each individual, and with every visit to the consultation room. It is an active process in which the patient feels understood and senses the attention and compassion of the doctor. Thus, the patient is identified by calling her by her name and invited to tell her story: 'How can I help you?' The doctor listens without interrupting, for what seems to many students a very long period. Here the body language of the doctor is attentive, eye contact with the patient is maintained throughout, while facial expressions, encouraging nods, hand gestures, and utterances of affirmation or surprise are important ways in which the patient realizes that the other is present and engaged. Now, however, I find making and maintaining eye contact much more difficult through the screen and the fine nuances of facial expression are lost on me in a myriad of pixels. I miss the privacy of a consulting room where I can purposely 'mute' myself and actively listen, and hear and see the story of the patient being shaped and thus bear witness to the sorrow and suffering. I am not sure about my screen presence either, as I am dressed for the occasion, wearing my white lab coat, and the rest of my attire, my body and face fixed in the proper camera caption, but I wonder whether I am actually present in the life of the other.[1]

A doctor seeks to elicit reliable information from the patient, in order to establish a meaningful representation of the complaints in her own mind. The doctor will summarize the issues that have been discussed and reflect what she has heard, seek confirmation of the patient, and deliberate what further actions are justified. In online communication, it is much more difficult to sieve through all the problems that are presented and more time-consuming to condense the relevant information.

The physical exam is a vital part of the doctor-patient interaction and is performed in a caring way, with attention to privacy during undressing, a proper examination couch, good lighting, an explanation of the steps taken in the examination, and explaining what has been observed. Handling the personal space surrounding each one of us in a caring and professional way must count as one of the most profound experiences in the patient-doctor relationship and also one that patients may remember much better than what was said. This does not mean that the stethoscope is going to be replaced by a magic or healing touch, but to be able to guide your hands comfortably, have a calming gaze and working in close proximity to your patient is a form of communication that is crucial. Now I am cut off from this clinical skill, not being able to establish contact and examine a patient and thereby losing an important way of creating a trustful relationship.

All of the above – attentive listening, explorative questioning, summarizing, reflective comments, eye contact, and actual physical contact in the context of a medical examination of the human body – is what will generate a healing therapeutic encounter that is conducive to improving the symptoms of patients, leading to reduced distress or disability. It is this drama of face-to-face interactions in a private space between a patient and a doctor that generates these so-called placebo effects, that are tangible, provide relief and sometimes dramatically improve the outcome.[2] The ultimate mission of medicine is to alleviate unnecessary suffering and in this I often find comfort, that in the absence of having a cure for a disease, I can generate, together with my patients, placebo effects that will ease some of the pain. Technocratic and biomedical means of intervention have a role to play as well, but it would be wise not to discard the potency of potential placebo effects.[3]

A good doctor, in sum, is able to develop and build a therapeutic relationship with the patients that seek their care. This relationship will benefit the patient, but it is also important for the doctor, and the job satisfaction of physicians is often better when they experience a good doctor-patient relationship. Medical education will provide for scientific knowledge and the training of clinical skills, but in order to raise and maintain a healthy workforce of caregivers it is increasingly noted that empathic medical care can be taught and trained in practice, and this has led to its incorporation into medical curricula.[4] As a doctor I feel privileged to be able to come very close to my patients and develop partnerships with them based on trust and compassion.[5] In the face of hardships patients go through, it is inspiring to witness their incredible strength, resilience, and the humour they muster and share. However, the weight of witnessing my patients' suffering may, at times, be overwhelming, especially when it resonates with own personal traumatic experiences.

I have pushed myself many times to go the extra mile in caring for my patients and forgot to set limitations, so it felt as if I were walking in their shoes. The blurring of these boundaries may consume one's own emotional resources, increasing the risk of burnout and compassion fatigue. The recognition hereof should lead to the mobilization of our own sources of resilience and to the gathering of support from colleagues, friends, and family or taking more time off. The next narrative is about that experience.

The last time I admitted Johnny was when he came in with a bout of pneumonia, as he had done several times before. Although he was coughing and short of breath he had walked into the ER on his own, in his high-top sneakers with loosened laces. Didn't your mother tell you to tie your laces before you go out? Because one day you will trip, I would say. I like high stakes living, he replied. The sneakers where a gift from his elder sister, while his younger sister had donated bone marrow to remedy his leukaemia. He loved them dearly as he clung to life. There was no doubt in my mind that our team and the medical might of the institution it symbolized would be able to give him a better life, and restore his joy of being. In the year that

I took care of him we became closer to each other. His father had left his family when he was eight years old and his mother did not look back in bitterness but instilled joy, recognition, and confidence in her children. He was a year younger than I and while I stumbled through my relationships, I admired his success with the nurses, who would flock to his bedside the moment he came in. Being the only man in our household refined my way with women, he said. Where would I be without them, as they have given me life in myriad ways and allowed me to stand on my own two feet, never doubting me. He would give me instructions on how to approach and fix a date with one of the nurses on the ward. We laughed about her rebuttal when she told me in no uncertain terms that I shouldn't let my imagined doctor status get the best of me. I lacked Johnny's elegant grace and buoyancy. I raged at his recurrent disease that he seemed to stoically accept, as well as the next round of therapies that gave hope, followed by disappointment. I need to go to the bathroom, he said to me, and I helped him from his bed, into his sneakers, and was about to tie his laces. Don't you dare, he whispered. I want to be able to step out the moment I want to. Oh, no, not on my shift, I said, as we inched our way to the bathroom. Halfway, he started coughing uncontrollably, his chest heaving and the infusion pump bleeping. Nurses rushed into the room and we helped him to his bed, increased his oxygen supply and were able to control his shortness of breath. I left him that night, quietly in his bed, with his shoes on his feet, as he had wished for. The next morning, I walked in, ready for a cheerful entry, and was struck by the sight of an empty bed with his silent sneakers standing next to it, side by side. Johnny had spent his last night in the IC unit and that is where he passed away early that morning. My legs felt heavy and I sat down staring at his worn-out shoes. Should have tied your laces, I thought, with tears rolling down my face. A nurse sat next to me. She put her hand on my hand and said, It's been a long journey and, yes, we walk together with our patients, step by step, never in their shoes, though, but close by, and there is strength in that.

A doctor will often need to balance the intellectual engagement and the emotional appeal that the ill or distressed patient may evoke. In order to provide compassionate care, we have to make space where this can be provided and claim time so it can be created, all the time for everyone, and in the meantime acknowledge our own needs as care providers and tap into our own sources of sustenance and support.

Caring well for others, in my view, is meeting the other in her loss or pain, witnessing the sorrow in her eyes, and reaching out by being present in that moment as a fellow human being, confirming a shared and fragile humanity. As a young man I was impressed by the human suffering portrayed in the book of Job, in the Old Testament, and it so resonated with societal problems like poverty and inequity, that I decided to enter medical school with the intention to put an end to human suffering and in the meantime uproot the system and improve the lives of as many people as possible. The medical curriculum provides students with the latest scientific discoveries of that magnificent contraption, the human body, and teaches you skills to examine patients, make a diagnosis, intervene with all kinds of means and even to operate on them when problems arise. These tools empowered me, and I eased into a vocation that suited me and strengthened my will to be of benefit to my patients, one after another, step by step. On the other hand, it increasingly dawned on me that employing powerful intrusions into peoples' lives, likewise came with the obligation not to cause harm. The bloodbath that developed when drawing blood from a patient on anticoagulant therapy or the dissatisfied patient who complained about the horror scar that I created after the excision of a mole, or the limping patient who ruptured his Achilles tendon because of the drug I prescribed for the urinary tract infection, are all readily recalled personal experiences. Many examples of iatrogenic harm to patients are hard-wired into the mental framework of physicians and other caregivers.[6] Finding a balance between these opposites of providing good care and not causing harm is not easy and cannot always be avoided, even with pure clinical reasoning based on scientific principles. Therefore, doctors are bound to make poor judgement calls when practising medicine, which eventually will harm some of their patients. We encourage doctors to be open about incidents and to share these stories, especially with our students, to create awareness, discuss preventive measures, show how to handle accountability in a professional way and hopefully gain redemption. The next narrative is about that experience.

He was a man in his early twenties who came under my care for a rare type of skin cancer for which we ran a specialist referral clinic. It took us a while to reach a conclusive diagnosis, and during these periods we would repeatedly examine the patient, replicate lab tests and biopsies from the lesions and provide symptomatic treatment. In the meantime, he and his relatives were well aware of the discussions that went on, and there was a great deal of stress and anxiety during the process of reaching a definitive conclusion. Eventually, we reached a firm diagnosis of a malignant lymphoma and, in an emotional visit, we discussed the potential options for further therapy. His disease was progressive and the cancer had spread, so it was imperative to administer potent chemotherapy as soon as possible.

This was a mode of therapy with which I was familiar and after discussing the expected outcome, duration, and the potential side effects, my patient started taking the prescribed medication. We regularly discussed side effects during subsequent visits, and I told him he was doing fine with the dose that I had prescribed and apart from regular blood tests no further action needed to be taken. The results were favourable and after several months of therapy I was able to taper off his medication and then stop it all together. He remained in remission and we ended up seeing each other for a follow-up visit about once or twice a year.

Several years after this episode he sent me a letter informing me that he and his girlfriend wanted to have children and had so far not succeeded in becoming pregnant. They had turned to a fertility clinic where they were informed that he was infertile and that the drugs he took for his lymphoma may well have played a negative role in this. He wanted to know my opinion on this matter and within a short period of time I found out that I had given him the proper dose, but that this dose was lethal to his sperm-producing stem cells. Before we started the chemotherapy, we had even discussed the cryopreservation of sperm cells, but I had reassured him, that with the low dose he was receiving, this would not be necessary. The next day I met with him and his girlfriend and explained that I had made the mistake of prescribing a drug that had caused his infertility and offered my apologies. They were devastated and in tears and could not comprehend that a doctor they had sincerely trusted over the years would let them down in such a profound way. I was in tears with them and said sorry and reported the incident to my hospital board. Here I had a hearing in which I took all responsibility for this calamity and an official complaint was filed by my patient. He was awarded compensation for damages by the hospital insurance company and he decided to seek care for his medical condition elsewhere.

I have used this calamity as a discussion and teaching case with my colleagues, residents, and students to show where my line of reasoning had gone awry. I usually worked with this particular drug in elderly patients, in whom this type of lymphoma is more common, and for whom fertility was not a major concern, and, thus not commonly discussed. Furthermore, I had calculated the safety margin for the cumulative dose with the figures of another drug in mind. It would have been as easy as just opening the drug insert and do the proper calculation to have come up with a different advice. Perhaps, back then, I thought I had taken long enough to reach a diagnosis and now, entering the therapeutic phase, I could not bear to seem hesitant about the course of action we needed to take and brushed away any doubts or proper calculations of the safety margins and pushed ahead without taking a moment to reflect. I plunged into a depression from which it took me a while to recover, while continuing my work in a more defensive way as my self-confidence was minimal.

Several years later I received an e-mail message from my patient. He wanted to share with me an image of his one-year-old son, whom he had together with his new girlfriend, who became pregnant with donor sperm from his younger brother. He was a proud father. I replied that years of sorrow and shame had not provided me with a way to undo the wrong I had caused him, but I congratulated him and said that I felt grateful that he had considered informing me about his son.

As medical professionals we are trained well in the workings of the body and what to do when things falter or go wrong. We have developed guidelines along which we proceed with diagnostic plans and therapeutic ladders. We document all our actions in electronic medical records that generate big data which, together with the massive amount of medical studies published each day, make some of us foretell that before long, our patients will be served by artificial intelligence and computational algorithms, which will make better diagnoses and treatment plans than we mortal physicians can.[7] Maybe this will work for some people, but for most of my patients this will not do. At times I say to a patient who may have a chronic condition, that 'I will make you better', because I know that, if done appropriately, the clinical encounter can truly become a healing encounter and that is what makes our profession at times very rewarding, not only for the patient but also for the care provider. I believe that a trusting doctor-patient relationship is paramount to a meaningful communication in an unequal relationship with a patient who may be suffering, sick, worried, looking for answers in an emotionally charged encounter with a health care provider who is taught to keep professional distance but has learnt to use their presence as a tool to alleviate suffering.

Suggestions for further Reading

H. Marsh. *Do No Harm: Stories of Life, Death and Brain Surgery*. London: Orion Books, 2014.

D. Ofri. *What Doctors Feel: How Emotions Affect the Practice of Medicine*. Boston, MA: Beacon Press, 2013.

A. Gawande. *Complications: A Surgeon's Notes on an Imperfect Science*. London: Profile Books, 2003.

LEVIATHAN

Or

THE MATTER, FORME and POWER of A COMMON-WEALTH ECCLESIASTICALL and CIVIL.

By THOMAS HOBBES of MALMESBVRY.

Law, imagination, and poetry

Bald de Vries

Right matters

> If right doesn't matter, if right doesn't matter, it doesn't matter how good
> the Constitution is – it doesn't matter how brilliant the framers were. [...]
> If right doesn't matter, we're lost. [...] If truth doesn't matter, we're lost.[1]

These are powerful words, spoken by senator (and lawyer) Adam Schiff
in his closing statement during the first impeachment trial of Donald
Trump. These words reflect a truth and they carry a message. There is
something at stake. The English poet Shelley observed in his essay *In
Defence of Poetry* (1840) that poets are the unrecognized legislators of
the world. But – perhaps at first sight far-fetched – lawyers might be the
unrecognized poets of the world. They might if they dare to be critical
and when they realize law is a piece of art, an artifice to be cherished and
looked after. Imagination can help in this respect, as part of legal practice
and legal education.

Our world is, again, threatened by autocracy. It is the cheap answer to
uncertainty in the complexity of contemporary global society. Autocracy
devalues law as if it were a mere decree. The decree of the ruler, the Leviathan, as an answer to the Hobbesian state of nature of a war of all against
all, where life of men is 'brutish, solitary, nasty and short'.[2] But uncertainty
is not 'solved' by autocracy. Rather, social uncertainty must be understood
for it to be transformed, enabling us to embrace it, while still being free as
individuals, and being a collective demos.

This essay explores the nature of law and how to teach it, by means of
imagination and poetry. It helps us to understand that law is not a mere
system of rules but a discursive practice based on normative foundations.
Imagination, allows us to explore avenues of moral reasoning to understand
law's moral foundations. It also enables us to empathize with people who
carry the brunt of the law or are involved in legal disputes. In the end,
imagination and moral reasoning contribute to developing an independent
mind with a sovereign voice.

Understanding law and society

We tend to think of law as a system of rules that prescribes what we can
and cannot do. And if we violate the law, a sanction will follow. So, to
understand law, is to understand the rules and how to apply them to the
case at hand. Easy. In doing so, we take a positivist approach, using the
model of subsumption, based on the case study method. We are presented
with a set of facts from which we distil a legal question that we answer by
reference to the law in the books, without critical reflection. James Boyd
White succinctly puts it like this:[3]

> The implied contract between the student and teacher shifts focus: our
> insistence to the student that 'You are responsible for these texts [the
> law in the books, BdV] as you have never been responsible for anything
> in your life', all too frequently entails the acceptance of a correlative as
> well: 'and responsible for nothing else in the world'.

Law is the result of dealing with societal dilemmas, for example the dilemma
of how we want to be governed. Law rests on conflict: law generates societal
conflicts in a particular way and in doing so provides the tools to address
these conflicts. We often present this as a technique: the problem is translated into a legal problem, or question, answered by the appropriate legal
rules: case resolved, the book can be closed. The 'solution' exists in deciding
along the scheme lawful-unlawful. As lawyers, we tend to neglect the scope
and nature of the problem and forget to imagine its societal context and
the people involved. It is an instrumental way of thinking, geared towards
solutions and measurable results.

But this is an oversimplification of law as a social and normative construct. It would be better to understand law as a hierarchical system of
different types of rules that govern social interaction between individuals
(e.g. citizens), organizations (e.g. companies), and institutions (e.g. the state).
It enables us to know what to expect from one another in a normative sense
and how to hold on to these normative expectations when they are not met.[4]
What I mean is that law governs those expectations that deserve protection
if violated or disappointed at all three levels – individual, organizational,
and institutional: theft, breach of contract, abuse of power, etc. And in
governing these expectations, law is more than a mere set of rules. Law
requires judgment. The simple reason for this is that in many cases legal
solutions do not simply follow from the facts – there is a 'gap' so to speak
between the law and the facts.

This implies, as the Dutch lawyer Paul Scholten shows, that a concrete legal relation does not merely depend on the rules but also on decisions. And these decisions cannot be found through the model of subsumption of rules (legal reasoning). Each time a decision is made, the realm of rules and reasoning is transcended by the realm of conscience (moral reasoning). Decisions are anchored in the judge's conscience from where he or she 'jumps' to the conclusion (the judicial decision), as a leap of faith. Scholten:

> I think that there is more than merely observation and logical argument in every scientific judgment, but in any case, the judicial judgment is more than that – it can never be reduced to those two. It is not a scientific proposition, but a declaration of will: this is how it should be. In the end it is a leap, just like any deed, any moral judgment is.[5]

The decision is an act. It nests in the judge's moral conscience for which he or she takes responsibility. Scholten stresses this point of responsibility, explaining that the judge's decision is perhaps not the only possibility that fits within the legal system and perhaps another judge would have come to a different decision, but for the judge, *this* judge, because it is a decision of conscience, the decision is the only option left, excluding all other options. But do we do justice, in our education and our programmes of study, to what it is to *judge* in this way?

What interests me, is how to think about judging as an independent act. In what follows, I focus on the relationship between judicial independence and professional and ethical attitude, and how to enshrine this in legal education. I think that the strength of legal education should lie in teaching legal consciousness: the ability to make judgments, forming opinions and taking a stance, and to acquire the (judicial) courage,[6] that comes with it. Using imagination is a way to explore avenues of moral reasoning with an aim to strengthen the ability to judge and to decide.

Imagination

Imagining is a way of thinking about how things are or ought to be. It is freer and more unencumbered, and different from academic (or legal) reasoning. Shelley, the great English poet, points to these two types of '*mental action*' – instrumental reason and imagination – in his essay *A Defence of Poetry*.[7] Reasoning, according to Shelley, is 'the act of contemplating the relations by one thought to another'.[8] It suggests a serial line of (logical) argumentation steps towards a particular conclusion. Imagining implies acting 'upon those thoughts, colouring them, leading to the composition of other thoughts', as an act of synthesis, rather than analysis, considering thoughts in their 'integral unity'.[9]

Shelley saw in poetry the highest form of imagining, so much so that without poetry, without poets, reason, science, and technological invention would not have flourished in the way they have.[10] Furthermore, in poets he saw the authors of morality and law. It is an elegant thought, and there is merit in Shelley's claim that 'poets are the unacknowledged legislators of the world', considering how we can be drawn to abstract poetic expressions of truth and beauty – to an 'image of life expressed in its eternal truth'.[11] Does his 'Declaration of Rights' not mirror the sentiment of Locke's Second Treatise (and warn us of the dangers of tyranny)?

> Man! Thou whose rights are here declared, be no longer forgetful of the loftiness of thy destination. Think of thy rights [...]
> Awake! – arise! – or be for ever fallen.[12]

Shelley's essay also points to something else. He suggests that imagination is a moral instrument for the improvement of man: 'It awakens and enlarges the mind itself by rendering it the receptacle of a thousand unapprehended combinations of thought. Poetry lifts the veil from the hidden beauty of the world, and makes familiar objects be as if they were not familiar'.[13]

Poetry as a means of moral improvement is perhaps too strong a claim to make, but undeniably, the mental act of imagining, which poetry is, opens up possibilities of thinking about the world (and law) differently – the world that exists is perceived and expressed in a universal metaphorical language, a world of which we are an integral part. In doing so, poetic acts can create 'new materials of knowledge and engender in the mind a desire to reproduce and arrange them according to a certain rhythm and order'.[14]

Using poetry to teach law

I had the opportunity to work with imagination and moral reasoning in a course I have been teaching at University College Utrecht for quite some time – Law, society, and justice. This course addresses some fundamental issues concerning law, power, and justice. At some point, students study the work of Thomas Hobbes and read Chapters XIII, XIV, and XVII of The Leviathan. It is not difficult to understand the central message: that a strong state is necessary to put an end to the 'war of every man against every man'. But what is the relevance of this theory? Is it necessarily true? Does the current political landscape seem to suggest this or are we misled? Students struggled with these questions as well as with the question why Hobbes is a *modern* thinker. It is the moment I introduce imagination, asking students to write a poem in which they connect the central message of Hobbes to developments or incidents in contemporary society. To help them think of this connection I refer the students to a 2012 column of John Gray (the English political philosopher and publicist), 'Hobbes, Our Great Contemporary'.[15]

Students have written some seventy poems over the past three years. Most of these clearly reflect, at least in my view, the themes Hobbes addressed in *The Leviathan*, indicating that students were able to 'translate' these themes into another idiom, while having an eye for contemporary life. Students referred to Putin or Erdogan, IS, the NSA and surveillance, even Mark Rutte (the Dutch prime minister) and also to Trump. A good many poems also stood out for their poetic quality. Here are three examples. (The copyright of the poems lies with the students, whose written permission I have obtained to feature their poems in this article, including the authorship.)

SMALL FISH
(Reinoud)

A man sits in a golden tower and foretells of greatness.
Strong he claims to protect us from all the horrors outside the gate.
Walls and work are promised in this Leviathan state.
He promises to release us from the boring and tasteless.

Fake hair and fake news, suddenly politics excites.
We all participate and everything's history in the making.
Yeah, that is the pretty truth, our new world is one of fights.
But are you sure it's not us that he's faking?

Listen closely to the great man talk so sweet,
It's small fish that sea monsters eat.

BRICKS
(Eva)

With building blocks, they run across the land to raise
A tremendously strong shield, to protect, secure.

The orders were to make cement, to stiffen
Their insecurities, to keep out what they fear.

This strength is what they envied,
what they wished to be guided by. So they keep building,
But their worries don't seem to ease.
Their insecurity increases, when they realise
Their control is lost, sucked up by this great wall
That was supposed to protect.

But now the bricks are up, and their freedom,
used as cement, is locked in their construction
They realize, there is no way left to go,
But for this narrow space, defined by the dictator
Of this great big wall.

ELITE
(Gheorghe)

They said that reason over passion rules,
Free world is dangerous, for us fools
We should be controlled, constrained
And liberty should not be treasured or obtained,

They gave a bread and called it meat,
They promised war will not repeat,
While all along we live on streets,
And sing our praise to the elite,
But now, secretly we read at night,
While praying to the lord in fading lights.

I started the experiment as a simple idea, as something really fun to do with
my students, who have an interdisciplinary focus; they obtain different
strands of disciplinary knowledge, attempting to connect the dots. This
experiment seemed to transcend the disciplinary approach, both integrat-
ing different ways of thinking and challenging students to be vulnerable
and to step out of their comfort zones. Perhaps we also became aware of
something else: our identity and responsibility as citizens and professionals,
be it as lawyers or otherwise. The experiment seemed to create a 'larger
community',[16] transcending discourses on law, philosophy, and literature

into something new.[17] The poems above also reflect that thinking about law (and power) involves thinking about real people who are subject to law and power, enabling a form of (professional) empathy.

But it did something more. Thinking in imaginary ways also opened up avenues of moral reasoning to address legal dilemmas.

Moral reasoning – from intuition to theory

In the English language, we refer to the judicial function as the administration of *justice*. It implies, I would argue, that this goes beyond the mere application of rules to facts. It involves a normative evaluation. This normative evaluation entails a particular approach towards moral reasoning.

One way of moral reasoning is explained by Michael Sandel in *Justice. What's the Right Thing to Do?*[18] In it, Sandel explores theories of justice by reference to all kinds of moral dilemmas. Each time, the central question is whether a just society should seek (through law) to promote the virtue of its citizens, their general welfare, or allow the law to be neutral about conceptions of virtue and general welfare and let citizens be free to choose the best way to live. Sandel engages the reader (and his students) with these dilemmas through a didactic of moral reasoning in four steps.[19]

The first step entails being confronted with a dilemma and trying to understand the dilemma. This involves, as the second step, an initial conviction or a 'conclusion tentatively formulated'.[20] Sandel then points out the importance of discovering the underlying principle of this conviction, embedded in theory. This requires, as the third step, the study of political philosophy, for example utilitarian theory (Mill, Bentham), Aristotelian theory, or Kantian ethics. Students discover that a principle underneath the conviction could lie in the notion of utility – promoting the general welfare of society for example, but then they discover, when changing the dilemma, we could get uncomfortable with our initial conviction and start questioning that principle. We discover for example that human life perhaps has some absolute value, worth protecting; that we can't use people as a means to an end. It is this all-important third step: a critical analysis of our initial conviction through philosophical investigation, which allows us to contrast and compare different ethical positions with an aim to formulate an ever-developing philosophical world view. And as an academic exercise, we can, as a fourth step, rethink our initial intuition, either by reaffirming or rejecting the original position, while being aware of the consequences. Nothing is at stake – true judgment is still 'suspended'.[21]

Through this exercise, we realize that once in a position of responsibility – when we actually *have* to decide, as a judge or in some other capacity, something *is* at stake. But for this process of investigation and discovery, Sandel says we need an interlocutor – to create dialogue.

Conclusion

What academic legal learning comes down to, is looking beyond the law and tying it in with other fields, such as political philosophy, in a way that helps us to understand the law differently. In other words, we can understand law (better) if we also study it *in the context* of other disciplines. Being confronted with a group of students whom I sought to engage in a discussion on law and power, reading Hobbes, the idea of poetry just came to mind, more or less out of the blue, as an implicit integration of thoughts I apparently had as to how to engage my students more meaningfully.

The 'trick' I used to bridge the two fields of knowledge – to integrate them – is to read philosophy differently, not as a lawyer but as an individual human mind, using poetry as a tool; to poetize or versify the essence of a philosophical text. It also paved the way for meaningful moral reasoning as a way to improve our ability to judge. It helps students (and not only students) to overcome uncertainty and insecurity and to discover a sovereign voice. Imagination contributes to the development of the reflexivity needed for this. Poetry can be used as a didactic tool for this pedagogy.

Suggestions for further reading

P.B. Shelley. 'A Defence of Poetry'. *Selected Poems and Prose.* Harmondsworth: Penguin Books, 2017, 635-665.
J.B. White. *Heracles' Bow: Essays on the Rhetoric and Poetics of the Law.* Seattle, WA: University of Washington Press, 1989.
K.N. Llewellyn. *The Bramble Bush: The Classic Lectures on the Law and Law School.* Oxford: Oxford University Press, 2008.

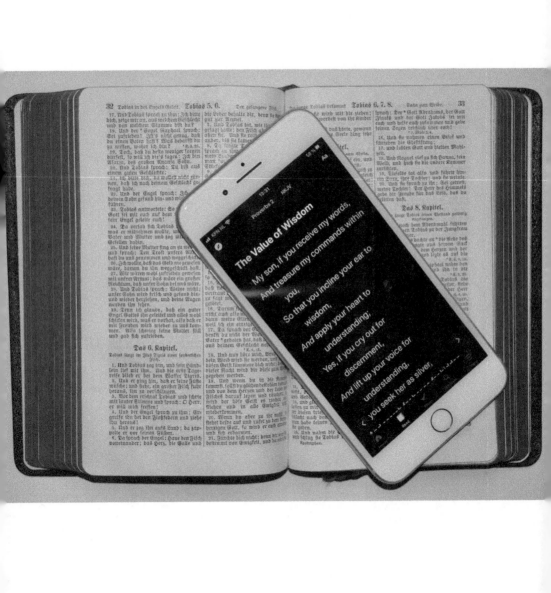

Religion 2.0: Thinking about religion through technology

Katja Rakow

Religion and modernity

We often perceive of religion as long-standing traditions reaching far back into the distant past and at odds with modernity. After all, the classical theories of modernization assumed that the more modern a society becomes, the less important religion would be in that society.[1] Indeed, the role and place of institutionalized religion in the West, especially Christianity, has changed since the mid-twentieth century. The decrease in church membership in Northwest Europe and North America is an indicator of the changing religious landscape.[2]

However, looking only at the Western world might skew our perception of the role and impact of religion in the 21st century. In most parts of the world, religion is alive and kicking. And even in the West, the assumption that religiosity is continuously decreasing has been proven wrong. Traditional institutional church membership might be dwindling, but alternative ways of being religious – might it be new religious movements, spirituality, or non-institutionalized forms of religious practices – exist next to more traditional forms of being religious.[3] Colonialism, globalization, and migration have brought religious traditions to new geographical settings. The rise of Christianity in South Korea or the establishment of Hinduism in the UK, where these traditions now form part of the local religious landscape, are a case in point.

Looking at the adoption of new technologies in religious contexts offers another interesting perspective on religion and modernity. Since 2015, Kofukuji, a century-old Buddhist temple in Japan, offers memorial services for robot pet dogs that are beyond repair. To help human owners cope with the loss of their beloved robot pets, temple priests use established funeral rituals and adapt those to robot dogs.[4] Kodaiji, a 400-year-old Zen Temple in Kyoto, has a robot priest named Mindar who preaches Buddhist sermons. The temple administration hopes that Mindar will help stir new interest in the age-old Buddhist tradition among the younger generation.[5] Today, many smartphone applications help religious practitioners in different traditions to include religious practices in their busy daily lives: Bible Apps help

Christians to engage daily with Bible texts; the Athan App provides digital calls to prayer for Muslims and also indicates the correct prayer direction by pointing the digital compass needle towards Mecca; in 2019, the Pope launched the Click To Pray App, which connects believers in their prayer.

The persistence of religion in the 21st century and its adjustment to digital culture and artificial intelligence demonstrates that religion and modernity are not necessarily at odds with each other. How can we explain the resilience of religion under the conditions of modernity? Before we can answer this question, we will first have to take a closer look at the concept of 'religion'. We will see that our conceptualization of religion influences how we make sense of religion in human history and the lives of human beings. Once we have come to a better understanding of religion, we will be able to see how religion transforms and how religious practitioners adapt their traditions to changing historical circumstances. We will use the relation between religion and technology to have a closer look at how religion adapts to new developments.

What do we talk about when we talk about religion?

For many years now, I have asked students enrolled in my 'Introduction to Religious Studies' courses during our first meeting, what they associate with the term 'religion'. The terms mentioned most often are usually 'belief' and 'God'. It indicates a shared common understanding of religion as something concerned with belief, more specifically belief in God (much less 'a god' or 'gods'). 'Belief' is not only mentioned first but also perceived as primary to which all other aspects of religion (worship, ethics, etc.) are secondary. Even though such a view is broadly shared, it is problematic in at least two ways.

The first problem concerns the genealogy of our understanding of religion. Although we use it as a term to describe a universal phenomenon, our notion of religion as belief (or faith) is the product of a specific historical development heavily influenced by (Protestant) Christianity. Belief has always played a central role in Christianity. But in Orthodox and Catholic Christianity, belief was embedded in a complex ritualistic and material structure (icons, images, incense, relics, the veneration of saints, processions, etc.). In the context of Protestant Christianity, faith has been elevated above the ritual and material dimension of religion. Think of the strong emphasis on the doctrine of *sola fide* ('by faith alone'), which distinguishes Protestantism from Catholic and Orthodox Christianity. In short, our notion of the universal category 'religion' is not so universal after all but firmly rooted in Protestant Christianity. It is thus not able to capture all dimensions of religion that are important if we want to understand the phenomenon.

The second problem concerns the assumption that belief is primary and everything else secondary, i.e. the idea that religious beliefs inform what religious people think and do. It is a top-down model of religion: if we want to know something about a religion, we study the beliefs and doctrines of a tradition codified in their religious scriptures. From these textual sources, often written by a small educated elite of (mostly male) religious specialists and literati, we transpose what the average religious person thinks, feels, and does. Such a model assumes that human beings function like automatons and do not have much agency to act on their own: religious people believe certain things, thus they act in a certain way.

Until a few decades ago, if historical records showed discrepancies between the official religious beliefs and practices (as encoded by learned religious elites) and the lived practice of religious laypeople (which is the majority of religious people), then scholars simply assumed that these people did not know any better. They were thus not worthwhile to be studied if one wanted to learn something about 'proper religion'. In the past, scholars of religion mainly studied religious texts and learned discourses and much less the actual practice of religious (lay) people. The common understanding of religion as something mainly concerned with matters of belief fitted well with such a scholarly programme. Although the common everyday notion of religion still centres on belief, the academic study of religion has moved on to a more nuanced and complex understanding of religion as lived and practised.[6]

From belief to practice

What happens if we think of religion not just as something that religious practitioners believe, but something that they do? If we make practice – things people do, how people act – the basis of our conceptualization of religion, then we see how religious practices are informed by beliefs, but simultaneously, how practices inform and shape the beliefs that people hold. A practice-centred view of religion has two advantages.

The first advantage is that we are able to perceive religion as something humans do in the context of their daily lives and their mundane living conditions, instead of thinking of religion as only or primarily concerned with the otherworldly dimension (i.e. the sacred, the hereafter). Admittedly, the otherworldly dimension is the content of religious beliefs, but how these beliefs are shaped, expressed, and experienced happens in the context of the mundane world human beings inhabit.

The second advantage is that it enables us to see religious practitioners – experts and laypeople alike – as active agents instead of passive recipients of the religious tradition to which they belong. As mentioned in the introduction, religious traditions are often perceived as a timeless, unchanging reservoir of practices and ideas, that gets passed on from generation to generation. What such a notion of tradition fails to acknowledge is that each generation has to interpret their tradition in the light of their own time and is thus actively negotiating what to keep, what to change, and what to consign to oblivion. In fact, the notion of tradition is often invoked to legitimize reinterpreted or new ideas and practices. Invoking the weight of tradition gives gravity to practices and ideas that people value now and which they want to see passed on to future generations: 'Notably, the idea that tradition reflects the past is misleading; the selective uses of tradition reflect what people want for their future'.[7] Foregrounding practice instead of beliefs allows us to think of religious practitioners as active agents who shape, interpret, and adapt their religious beliefs and practices to their own time and circumstances. After all, religion itself does nothing. It is religious practitioners who do things with religion, to religion, and in the name of religion. If we want to know something about religion, we have to study what people *do* with religion and not only what people believe.

With such a dynamic understanding of religion – an understanding in which people are shaped by beliefs and practices and actively shape beliefs and practices – we are much better equipped to assess the resilience of religious traditions through the ages. It will help us to perceive religions not as static and unchanging traditions, but as complex figurations of practices and beliefs. Religion is not independent of but entwined with countless other spheres of human discourses and practices such as politics, economy, education, medicine, art, media, etc. Changing historical and cultural settings thus impacts religious thought and practice and *vice versa*.

The transformation and continuation of religion

The invention of new media and technologies provides an interesting lens to explore how religious traditions adapt, transform, and thus continue to exist under new conditions, albeit in sometimes subtly changed, sometimes radically altered form. We will use the example of Christianity to trace some intrareligious developments that have been shaped by the implementation of specific extra-religious practices and technologies.

The printing revolution

The invention of the printing press in fifteenth-century Europe and its effects on Christian practices, especially on reading the Bible in vernacular languages, is widely acknowledged. Even before translating Bible scriptures into German, Martin Luther paved the way for the mass distribution of religious pamphlets to the common people so characteristic of the entwinement of book printing and the spread of the Protestant Reformation. In the beginning, Luther's disputes with the Catholic Church were first published in Latin, but he soon realized a growing interest in these debates beyond clerical circles. Five years after the public display of his 95 theses, almost 90 per cent of his writings were published in German.[8] Besides Luther being a prolific writer, the format of his religious pamphlets also played a significant role in the successful spread of his ideas. Luther's texts were printed in the easy-to-produce and cheap-to-acquire quarto format (singular sheets, which were folded twice to produce four leaves, i.e. eight pages). The quarto became the characteristic format for many of the Reformation texts and helped to spread and circulate its theological debates widely. These early Reformation pamphlets often included an explicit request to share the reading with the illiterate, and the text was formulated in a way that it could be read aloud easily. Luther's skilful feeding of the growing demands of the reading public and his use of commercial printers, the quarto format, and vernacular language could be described as 'the first mass propaganda campaign'.[9]

His Catholic opponents stuck to Latin as the language of theological dispute and the church as a designated site for such debates. By eschewing German as a suitable language and the marketplace of cheap commercial printing as a suitable site to address Luther's challenges, the Catholic Church had a difficult time in successfully countering Luther's efforts in the early stages of the Reformation. It was not until the second half of the sixteenth century that the Church was trying to oppose Reformation efforts more proactively. But by that time, 'vernacular Protestant versions of Christianity' had taken hold in several European countries.[10] In these strongholds of the Reformation, the language of faith had been irrevocably changed. Luther and other reformers were early adopters of a new technology, which had a huge impact on the formation of a new strand of Christianity and wider cultural practices. Protestantism's strong emphasis on religious reading led to the advent of individual (silent) reading practices and with it a noteworthy increase in literacy within the broader population.[11] Both developments significantly shaped modernity. To date, the reading of printed texts is still one of the key cultural practices to access information about almost everything.

The digital revolution

The introduction of digital technologies and the widespread use of handheld devices such as smartphones had a huge impact on everyday life and our working routines. It is no surprise that the digital revolution also had an impact on religious practices. While Christian publishers jumped on the bandwagon of digital publishing early on and announced the triumph of the digital Bible, others opened the discussion of the appropriateness of digital Bibles in religious practices. We might assume that it makes no difference as both versions – the printed and the digital Bible – contain the biblical text. But a closer look at the ongoing negotiations among Christians about the suitability of different Bible versions shows that it depends on what religious actors intend to *do* with the material object called the Bible. If Christians want to read and study the Bible text, they might prefer a digital version (online or a Bible app) because of the many practical advantages. Twenty years ago, for a comparison of Bible verses in different translations, the reader would have needed to look into different bound volumes. She would have needed those books at hand as well as time to flip through the many pages, to look up the different versions, and to put them next to each other. Digital Bibles not only contain many different translations and editions of the text, but also supplementary material that enriches the Bible text for further studies (e.g. different annotations, maps, commentaries, and other explanatory resources). Digital Bibles overcome the limitations of physical books when it comes to reading practices, Bible study, exegetical purposes, or preaching. The option to enlarge text lettering comes in handy, especially for usage in the pulpit. Younger Protestant Christians and newly converted Evangelicals especially appreciate Bible apps as they help establish a daily reading routine through push notifications and structured thematic reading plans. Some of them pointed out that they get lost in the printed book and don't know where to start.[12] Unfamiliarity with the Bible thus can become an obstacle to fulfilling the Protestant imperative to regularly read the Word of God. For some Christians, Bible apps are a supportive medium to learn to navigate the Bible and develop their devotional reading practice.

The advantages of the digital Bible for devotional reading and study might turn into a disadvantage when it comes to liturgical and ritual practices. Catholic Christians might have a hard time imagining their parish priest processing an iPad down the aisle and placing it on the altar instead of a Bible book. In Pentecostal healing services, the minister might place his Bible on the head of a sick member of his congregation attempting to drive out the sickness while invoking Jesus' name. Again, he and his congregation might struggle to imagine replacing this with an iPad on the believer's head.

Like the printing revolution that allowed for new and more individual ways of engaging with religious texts, the digital revolution makes the Bible more accessible for current generations, creates new ways of spreading the Gospel, and results in innovative religious practices. For now, the printed version coexists with the digital version, and which Bible version is chosen depends on what people find most suitable and practical for their purposes. What the long-term effects of the digital revolution might be for cultural practices in general and religious practices more specifically remains to be seen.

Conclusion

The importance of studying how religions adopt new media and technologies lies in their relevance for the transformation, and thereby continuation, of religious ideas and practices.[13] Religious actors might resist technological change and see it as a threat to established religious authority, practices, and beliefs. Yet, religious practitioners have to take a stance concerning these new technologies – might it be (partial) adaption or (total) refusal – if they want their religion to stay relevant to their adherents. Even if these technologies are officially banned from religious contexts these technologies may still impact the daily lives of religious practitioners. Whether religious actors enthusiastically champion new technologies or fervently argue against them, for both positions, they will invoke tradition and selectively use the same textual sources to legitimize their views.[14] To understand these complex processes, it is not sufficient to only study what religious practitioners believe, but also how they practice.

We have come full circle. We can explain the resilience of religion through the active efforts of religious practitioners to sustain their traditions. They actualize their religion in the context of the particular challenges of their time. Only history can tell if the inventive efforts of 'early adopters' of new technologies like Luther in the sixteenth century or the Buddhist priests at Kodaiji-temple in the early 21st century have the power to profoundly reshape religious practices and discourses in the long run.

As we saw, religions are not timeless, unchanging traditions, but malleable and complex figurations of practices and beliefs. If we study religion not only in terms of belief, but as something that is practised, then we are able to account for the entanglement of religion with all other spheres of human activity (for instance, politics, economy, media, etc.). We will see how religion is shaped by cultural practices and technological developments, but also how religion impacts cultural practices (such as the advent of individualized reading practices stimulated by Protestantism). We have to venture beyond the narrow confines of studying religious texts and beliefs and take a cue from sociology, anthropology, history, media studies, science & technology studies, and others to get the full picture.

Suggestions for further reading

J. Stolow. 'Technology'. D. Morgan, Ed., *Key Words in Religion, Media and Culture.* Milton Park: Routledge, 2008, 187-97.

M. McGuire. 'Sacred Place and Sacred Power: Conceptual Boundaries and the Marginalization of Religious Practices'. P. Beyer, L.G. Beaman, Eds., *Religion, Globalization and Culture.* Leiden: Brill, 2007, 55-77.

K. Rakow. 'The Bible in the Digital Age: Negotiating the Limits of "Bibleness" of Different Bible Media'. M. Opas, A. Haapalainen, Eds., *Christianity and the Limits of Materiality.* London: Bloomsbury, 2017, 101-121.

Global mental health and the evolution of clinical psychology

Robert Dunn

Clinical psychology is moving in a very new direction as the second decade of the 21st century comes to a close. We have become increasingly aware of the fact that mental illness occurs around the globe, but there are simply too few people to provide help for those in need. A new field has emerged in the past two decades that aims to address this problem: Global Mental Health. Researchers are seeking to better understand the prevalence, impact, and effective treatments for mental disorders around the world. People from the fields of psychology, psychiatry, medicine, anthropology, public health, and even economics are collaborating, in hopes of speaking to this important problem.

Common mental disorders (CMDs), which include depression, anxiety, and post-traumatic stress disorder (PTSD), are a leading cause of disability across the world. It is estimated that depression will be the primary burden of disease by 2030. Globally, an average of between 4 and 6 per cent of any population, in any country, experiences depression at any given moment in time, and between 5 and 7 per cent are suffering from anxiety. It is estimated that twelve-month prevalence rates for all CMDs worldwide is 17.6 per cent.[1]

Only in recent years have we begun to understand that a clear relationship exists between physical and mental health. Mental disorders increase the risk of both communicable and non-communicable disease, and disease increases the risk of mental illness. Mental disorders not only impact a person's well-being, but also represent a considerable cost to society. If we can find ways to effectively reach out and treat those who are suffering, this will affect the economies of whole nations, not only the well-being of its citizens. The World Health Organization has fostered awareness of this nascent field of Global Mental Health by declaring 'mental well-being is a fundamental component of WHO's definition of health', and is developing 'action plans' in an ongoing effort to implement worldwide changes in mental health treatment.[2]

In high-income Western countries, effective methods have been developed to successfully treat CMDs. Cognitive behaviour and interpersonal therapies, among others, have been shown to be highly effective. But there is a problem with the lack of sufficient resources to provide adequate treatment. In the

Netherlands, for example, there are 20.8 psychiatrists per 100,000 in the population. Numbers are similar for practising clinical psychologists. This clearly is insufficient, though, as the waiting lists are typically three to six months to schedule a visit.[3] Low and middle-income countries typically have far fewer mental health professionals. Worldwide prevalence of psychiatrists is 1.27 per 100,000. In many low-income countries that number is less than 0.1.[4] Treatment, then, is reserved only for the most severe cases. A far greater number of people who are suffering, are not receiving help that would benefit overall health and well-being.

Researchers and clinicians are addressing this 'treatment gap.' New models are being developed and deployed whereby people can benefit from the knowledge gained by work already done. Local non-specialists (nurses, aides, community health aides, social workers, and other paraprofessionals) are currently being trained in many countries to deliver psychological help for those suffering from CMDs, and are quite successful in doing so. This has been seen in Zimbabwe, where grandmothers are trained in delivering short-term therapy problem-solving therapy (PST), and demonstrating striking success.[5] It is seen in Uganda and Zambia with the Strong Minds Programme, where group interpersonal therapy is delivered by trained members of the community. They have treated 70,000 people and found significant reductions in depression, and increases in income, in family food security, and even in their children's school attendance.[6] Clinical psychologists working collaboratively with community members have found ways to align the content of treatment with local traditional beliefs about health and illness. By offering treatment in culturally appropriate and accessible settings and by training people to simply be empathic, depression lessens.

Clinical psychology, then, is broadening its scope. Rather than limiting treatments to the well-off in the high-income countries, it is now seeking to develop culturally appropriate treatments with paraprofessionals around the world to do the therapy and, in so doing, helping many more people. This exchange will likely have a huge impact on world health, thus providing social, psychological, and economic benefits all around.

Case study – Jamaica

Jamaica is the third biggest island in the Caribbean, the largest English-speaking one, with a population of some 2.9 million inhabitants. It is about one fourth of the total landmass of the Netherlands. Nearly 80 per cent of the people are of African descent, having been brought over as enslaved people from West Africa during the sixteenth through eighteenth centuries. The people largely hail from the Akan Ashanti, Yoruba, and Ibibio people of Ghana and Nigeria. During their decades of enslavement, family structure, cultural traditions, systems of health, healing, and spirituality were quashed by their originally Spanish, then British captors. Though slavery was formally abolished in 1834, independence from the British did not occur until 1962. Since that time, an autonomous Commonwealth nation has developed that maintains allegiances to both its African heritage and more recent British structure of governance, educational orientation, and politico-religious traditions.[7]

The health-care system in Jamaica is comprised of two sectors – public and private. The public health sector is financed by the government. All Jamaicans have access to free or very low-fee health care. A total of some 375 health centres and 24 hospitals are distributed throughout the island, and some 90 per cent of the population lives within ten miles of a health facility. Most people utilize public, rather than private, health care.[7]

This health-care delivery structure is known as a primary health model. Promoted by the World Health Organization in 1978, the notion is to provide readily accessible services to people close to where they live and work. This has been of tremendous benefit to people in low- and middle-income countries (LMICs). Both communicable and non-communicable diseases can be addressed far sooner and at lower cost.

In 2001, the WHO set forth ten recommendations to reduce the treatment gap in mental health services.[8] The first recommendation is to include mental health services in primary care. This model was, in fact, implemented in Jamaica, whereby physicians, nurses, and paraprofessionals have been trained in detecting symptoms of mental illness with the aim of improved treatment, follow-up, better overall health outcomes, and a more efficient use of very limited human resources. In Jamaica, health professionals have developed a model known as 'Integrated Collaborative Model of Care'. The goal is to integrate mental health care into the general health system. The endeavour has met with some success, albeit limited. A major positive result has been a shift in placing patients in general medical wards rather than separate housing in mental hospitals. The main psychiatric hospital in Kingston has reduced its population from 3,000 in 1970 to 800 today. This has resulted in shorter hospital stays, better involvement with families in treatment, and more comprehensive reintegration of patients back into their communities.[9]

Depression screening was introduced into the primary health-care services in 1994. Though this was successful in identifying depressed patients, services remained quite limited. Though 6 per cent of the overall Jamaican health budget is spent on mental health, which is relatively high in an LMIC, this only allows for limited services for the many people in need. Only the more severe cases are referred to psychiatric treatment. This only increases the sense of stigma attached to mental problems. If such cases are the ones receiving treatment, folks with problems of less severity don't want to be seen to be in need of help. Still, increased sensitivity to patient's emotional distress and its impact on general health has led to better treatment than was the case just two decades ago. Yet more needs to be done.

Research project

A pilot study has started, a collaboration between the University of the West Indies and me, aiming to address the treatment gap and provide people who are depressed with short-term therapy given by trained paraprofessionals. Similar projects have been set up in low-income countries around the globe: among them, Zimbabwe, Uganda, Bangladesh, Chile, and India.

The clinical research project takes place at a primary care setting. This particular facility provides the range of health care services for August Town, a low-income, high-crime neighbourhood in Kingston. On average some 1,300 patients are seen to there every month, people with chronic diseases, such as diabetes, cancer, heart disease, stroke, and cerebrovascular disease. They are seen for the range of acute problems as well, ranging from such events as broken bones, influenza, unintentional injuries, or an asthma attack. There is a perinatal and a well-baby clinic as well. Our mental health team of four psychologists and a psychiatrist has trained high-school educated Community Health Aides, the people who greet and sign-in patients coming in to see a doctor, to conduct a six-session 'Problem Solving Therapy'. This therapy was developed by Mynors-Wallis and has been adapted for use in other LMICs (Mynors-Wallis, 2001).[10] We adapted it for Jamaicans.

Depression is very stigmatized in Jamaica, as it is in most of the world. In this project, the word 'depression' is never used when presenting it to patients. Instead, they are introduced to a 'health improvement programme'. People are told that they may be eligible to enrol in a project shown in many parts of the world to improve overall health and well-being. They then are given a cross-culturally validated survey assessing the presence of depressive symptoms (PHQ-9, the Patient Health Questionnaire, a quantitative scale measuring symptoms of depression). If they show symptoms that suggest mild to moderate depression, they are offered the treatment. A focus on improving overall health, then, provides a tangible, and non-stigmatized, reason to participate.

Community Health Aides (CHAs) implement a very structured short-term psychotherapy. The general aim of this treatment is to help a person feel more empowered to take on daily problems. Quite commonly, those suffering from depression are readily overwhelmed by the difficulties they face and feel ineffective and hopeless about achieving any measure of change. The approach of the therapy involves breaking down concerns and setting forth a 'SMART' goal to be achieved each week between sessions: a behavioural objective that is Specific, Measurable, Achievable, Realistic, and Time Limited. This is a cognitive-behavioural therapy approach, one that has in recent years been shown to be very effective in reducing depressive symptoms for people in Asia, Africa, and South America.

The project is in its early stages. CHAs have been thoroughly trained and patients have been seen. Their work is supervised weekly by psychologists. Ongoing training occurs, giving workers the confidence and knowledge to develop their skills. The present Covid-19 outbreak has, however, required a temporary interruption of the project.

Mental illness and culture

Many interesting issues arise when translating methods of psychotherapy that have been shown to be effective in Western, high-income countries to LMICs. Of central importance is the idea that culture very much shapes our identity. Our way of life, traditions, and values, our social identity, is located within a particular culture. So, too, culture influences how we understand illness, distress, healing, and the return to normal functioning. It provides systems of meaning and explanatory models to help make sense of illness, disease, and distress. Suffering itself is understood and expressed in a way unique to culture.

A person's understanding of how a psychological disturbance originates is very much based upon the language of illness, the 'cultural idiom of distress'. Prevailing illness beliefs are quite varied in different parts of the world. In one culture illness might be understood as resulting from a transgression of an ancestor, in another as spirit possession, yet another the workings of a rival having cast a spell. Arguably, our own Western explanatory model of mental illness is shaped by current cultural paradigms reflecting biologically based diagnostic systems involving systems of the brain, genes, and epigenetics.

This is very much shaped by cultural values honouring the primacy of scientific fact. In addition, our bio-psycho-social model, a prominent Western worldview in recent decades, suggests a cultural understanding of illness as the misalignment of the individual in relationship to oneself, or one's interpersonal world. Illness is understood as an individual, rather than a collective, problem.

In many parts of the world, bodily symptoms are the most common clinical presentation of a mental disorder. Pain, tiredness, fatigue, specific or generalized symptoms of malaise are often the way in which common and more significant psychological disorders present themselves. Focus on either somatic, emotional, or cognitive symptoms varies by culture; our worldview shapes and becomes the means of articulating our expression of distress.

This has clear implications for treatment. How I understand the cause of my problem relates to my understanding of the course of the illness and the type of treatment required to achieve a state of health. Making offerings to gods, healing ceremonies, or herbal treatments are employed in one culture in the same fashion as 'talk therapies' are employed in other cultures. Can it be, then, that treatment for what we in the West have found to be effective for what we have labelled 'depression' be effective in treating those in cultures with fundamentally different explanatory models of illness and health? The answer at this relatively early stage of global research is a qualified 'yes'.

We are exploring this question in Jamaica and researchers are investigating it elsewhere. In the Caribbean, ties to African religious tradition existed long before the introduction of Western biomedical approaches to health and illness. An Afrocentric explanatory model, involving beliefs in spirits, God, ancestors, and sorcery has a long-standing tradition in Jamaica. Herbal remedies have been commonly used to treat a variety of somatic and psychological disturbances for centuries; knowledge has been passed down largely through oral tradition. Healers have been important to people. Known as 'Obeah', they have occupied a vital place as practitioners of traditional medicine for many decades. The exact origins of this healing practice are unclear, but it is generally agreed that they are rooted in traditions of West African culture (likely the Igbo, Akan, or Efik people). Since their arrival in Jamaica, people have employed the services of Obeah for both healing and more sinister purposes. Health-care provision on the plantations was limited during the period of slavery, and people maintained health-care practices based on their original cultural values.

Through the decades, though, a European biomedical model of health and illness gained ground. Increasing numbers of doctors, nurses, and health-care professionals were trained in a Western model, a paradigm with different conceptualizations of mental illness and modes of treatment. These quite distinct worldviews, the Afrocentric and the Eurocentric, came to a critical juncture in 1898, when the British-run colonial government outlawed the practice of 'Obeah'. Declaring it a pagan and superstitious belief system, having no basis in 'science', those in power decided it was in the best interest of civilizing the by-then freed slaves to adopt a Eurocentric paradigm, whereby the doctor was the best person to treat illness, not the Obeah. 'Anyone who uses or pretends to use supernatural powers or knowledge is liable to imprisonment', the act stated.[11] Traditional healers were criminalized.

Now, more than a century later, Obeah is quite 'underground'. Few openly advertise their services, and it is reportedly only more commonly seen in rural mountain areas than in towns and cities. But local psychiatrists and psychologists report that patients' understandings of the more serious disturbances are often rooted in traditional belief systems. Dre, for example, a patient hospitalized with a diagnosis of schizophrenia: 'The doctors say it was hallucination, but I know I wasn't hallucinating, for me that was an encounter with some supernatural stuff'. Ariana, diagnosed with a psychosis, comments about her disturbance: 'I think witchcraft is a part of it, I don't know what part of it is necromancy [...] witchcraft can start mental illness'.[12]

It is as yet unclear how people understand the causes for depression or anxiety in Jamaica. Symptom presentation is quite similar to what people experience in high-income countries. Our research project does not explore what patients' understandings of depression are, the word depression is never employed. We simply look whether they show symptoms, and if so, offer treatment. The hypothesis is that a breakdown of problem-solving skills, and accompanying feelings of being overwhelmed by life difficulties underlies the depression. And if problem-solving skills can be taught, depression will remit. A change in thinking will lead to change in behaviours, and ultimately a change in mood.

If we see positive results from the treatment, many questions yet remain as to what brought about the change. Is it the very content of the method, the cognitive and behavioural activation techniques of the six-session model that makes for change, or could it be something else? Is it that the people who deliver the treatment have commonality and understand local 'idioms of distress'? Clinical research is increasingly focused on what are considered the 'non-specific' elements that seem essential to all successful treatments. These involve aspects of the therapeutic interaction that are not specific to any particular therapy, but rather common to all. The act of engaging with another person, of collaborating, and of conveying empathy appear quite important elements to therapeutic effectiveness. The trained caregiver who actively listens to what someone is saying and conveys that they are hearing what one is feeling, seems essential to successful therapies. Particular techniques (interpersonal, emotion-focused, cognitive or behavioural) have certainly shown effectiveness, but the importance of elements common to all suggests that more advanced therapies are yet to be developed.

Clinical psychology, then, is in a stage of essential growth. Addressing the mental health needs of the world's population is important, and the model that has developed in high-income countries is simply not easily translated to other cultures. We need to find economically viable ways to scale up services so that the health needs of greater numbers are met. CMD-treatment content empirically supported by research needs to be well aligned with local illness beliefs. This requires the collaboration between professionals from high-income countries where culture-specific treatment methods have been shown to be successful, and professional and non-specialist providers in countries where such methods are desperately needed.

Suggestions for further reading

A.E. Becker, A. Kleinman. 'Mental Health and the Global Agenda'. *New England Journal of Medicine*, 213 (2013), 369, 66-73.

D.R. Singla, B.A. Kohrt, L.K. Murray, A. Anand, B.F. Chorpita, V. Patel. 'Psychological Treatments for the World: Lessons from Low – and Middle – Income Countries'. *Annual Review of Clinical Psychology*, 13 (2017), 149-181.

V. Patel, H. Minas, A. Cohen, M.J. Prince, Eds. *Global Mental Health: Principle and Practice*. Oxford: Oxford University Press, 2014.

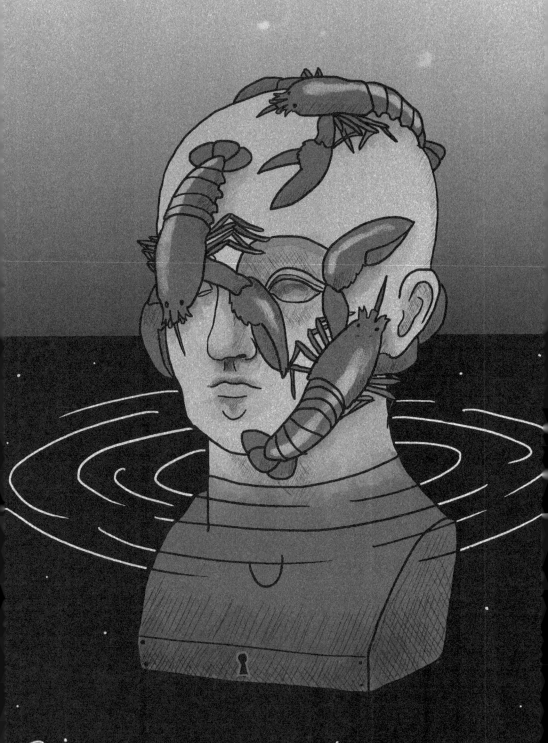

Everything makes more sense now

Heroes of the in-between

Rozi Tóth and Gerard van der Ree

Personal Narrative #1: Gerard

February 2019. It is 21 degrees Celsius outside. In February! I leave my coat in the office and go out for a walk. Everywhere students are sitting outside in the grass enjoying the sun in short-sleeved tops. Sunglasses abound. The sound of chatter resonates around the campus. Within twenty minutes I have the same conversation four times. Students tell me: 'This weather is so nice! I am enjoying this so much!' And then a silence, after which suddenly, with a worried look on their face, they continue: 'It is a bit scary, though.'

Suddenly a feeling catches up with me, and it lingers around the whole campus. *Climate anxiety*. A distant feeling of dread for something that is far away, but close at the same time. It does not feel like the fear of a clear and present danger. It is also not merely anxiety, which I understand as the experience of fear in the absence of threat. It is something more subtle. Climate anxiety seems to hang low, underneath other emotions, under the radar. It darkens thoughts about the future, even when it remains unsaid or even unthought.

I begin to understand climate anxiety as the fear of a world collapsing. Like fear of an earthquake. Usually when we are afraid, we fear things that are moving *in* the world. The world itself remains stable, which helps us locate the danger and define our responses. But what do you do when the world itself starts moving, starts threatening us? How do you keep your balance when the earth moves beneath your feet? I sense the fear. It feels like the beginning of a world ending.

What does it mean for the world to end?

Being bombarded with news, images, and stories of climate change, it is hard to stay optimistic about the future. The technical language of 'tipping points', 'feedback loops', and 'rapid loss of biodiversity' all points towards a dark horizon: the end of the world. But what would that mean? In Western culture, the 'end of the world' typically comes with the Christian imaginary of an apocalypse: a period of violence, chaos, and disorder, after which life on earth, and even earth itself, comes to an end. The end of the world, in that imaginary, is the end of the earth with everything on it.

However, the 'world' is not the same as the 'planet'. A planet like earth is a set of material things and beings. Yet, a world is always something more than those material things. It is a place of meaning, where the parts of which are held together so that they become intelligible, meaningful, and valuable. Rather than being a collection of things, a world is where meaning arises through experiences.[1] Our planet is not going to end. It will merely change. But our world – the world as we know it – may well end.[2]

Personal Narrative #2: Gerard

Greta Thunberg's 2019 'How dare you!' speech to the UN shocked me. Her accusations, that we – my generation – are not doing what should be done, that our words are empty, that we have failed the generations of the future, all of that fell heavily on me. Because it felt true.

I was born in 1968, only a few years before the Club of Rome report came out. That report clearly showed that our way of life was unsustainable. And in decade after decade, the evidence amassed: if we do not make dramatic shifts, humanity, and all life on earth, is going to be in deep, deep trouble.

Yet even as I grew up with these warnings, I did not respond. I lived my life in knowing denial. Yes, I ended up insulating my house and separating my garbage. But otherwise I was able, like so many of my generation, to keep the chimes of doom at a distance for a long time, and to pretend that a crisis was not unfolding under our watch.

Greta Thunberg is right. It was me, and my generation, who dropped the ball. And simultaneously, it is the people from my generation who are now in places of power. How do I respond to this without getting stuck in feeling guilty? I may not be responsible for the whole situation we are in, but that does not mean I do not have a responsibility. How do I respond? Where do I begin to connect to those who come after me, in such a way that all our voices equally contribute? How can we learn to respond together to a world that is – at least as we know it – ending?

If the world is ending, it is a world in which technology and capitalism determine a lot of what human connections can be. It is a world in which capitalism, industrial growth, and exploitation of natural resources play a dominant role, shaping the relations between humans and nature through instrumental logic. Seen from this perspective nature became a clockwork, whose regularity is a perfect background for human action. There was no agency to be expected from nature, no surprises, no unexpected responses. Most humans could live in disconnect to nature, unearthly.[3]

The climate crisis has begun to shift this frame of mind. It reveals the current 'world age' to be the *Anthropocene*; the geological era in which the natural world is primarily shaped by human activity. In the Anthropocene, humans determine to a large degree what the earth looks like, and simultaneously, it shows the earth responding to this human interference. The Anthropocene reveals humans and nature to be deeply entangled. Through this entanglement, nature becomes a threat to human existence. Simultaneously it is an invitation to humanity to learn to live with – and not against – nature.[4]

A beginning after the end?

The end does not have to be the end. It is a particularly Western (and Christian) conception of time that offers us a clear beginning and a clear end.[5] However, in many non-Western cultures time is understood to be more circular, offering new beginnings after each end. Facing the end of the world, then, does not have to mean facing the last stretch of existence. The end of the world can also be the end of a culture, of a way of being, of a civilizational order, which has been giving meaning to the people who inhabit it. If we understand the 'end of the world' in terms of the end of a civilization, or of a culture, then in the Anthropocene a way of living is reaching its end.[6] This end may well come with chaos, disorder, and violence but it does not necessarily imply the end of all life. The world as we know it may vanish, but a new world will become available in the process.

If the Anthropocene reveals planet earth as moving, acting, and responding to humanity's actions, it challenges our imaginaries, institutions, practices, and sources of agency. Currently agency comes from institutionalized political action and the shifting of consumption behaviour. As citizens we can contribute by participating in politics through voting, being active in political parties, and protesting. We can raise awareness about climate change and support political groups and civil society organizations that promote a more sustainable future. As consumers we can take responsibility by reducing our global footprint, flying less, eating less meat, buying second-hand clothing, and so on. All of this needs to be done. However, we cannot rely on these actions alone. Our political and economic structures are part of the problem that we are facing. They are therefore limited in responding to challenges that include their own transformation. We have to learn how to complement them with alternative and more earthly imaginaries, institutions, and practices that open up new forms of agency.[7] Responsibility then, as Donna Haraway writes, lies in the ability to respond.[8] We must become response-able to worlds ending, to ways of life dying, and to new worlds and other ways of life being born.

Learning to die

Part of the ability to respond is what Roy Scranton calls 'learning to die'. Learning to die is to look our deepest fears in the eyes and stop avoiding the inevitable. This does not mean despairing, though, or entering into a state of panic. Scranton says that for an individual, learning to die means 'letting go of our predispositions and fear'. But at a collective level, learning to die means letting go of 'this particular way of life and its ideas of identity, freedom, success, and progress'.[9]

Learning to die, then, means letting go of the things that make us who we are. We have to let go of a world we have come to know as ours. We have to let go of a way of life that is *our* way of life. We have to let go of the future, of our hopes and expectations. We have to let go of our identities, of who we think we are in this world, and what it is that we have to do there.

> **Personal Narrative #3: Rozi**
> I always had an image of the house I wanted to live in. One whose wall-sized, steel-framed windows look at a lake and the surrounding mountains, maybe somewhere in Switzerland. The interior is clean, made of wood, glass, marble, and concrete. The ground floor has an inside swimming pool and there is a small botanical garden with a view on the mountains. This house represents the future I aspire to; solid, carefully designed, and environmentally sound. A home for the future me.
> This summer I realized I am never going to build this house. It dawned on me that I live in a time of climate change so I cannot afford this future. The loss of this house and the future it represents, made me sad, confused, and slightly lost, emotions from which I urgently needed to find a way out.

The emotion that comes with letting go is grief and mourning. It is working through the emotions of loss and bereavement, not only as individuals; but also as a community. We have to put grief on our public agenda, and find rituals, forms of art, and collective language for saying goodbye to that which has to go. This is one of the ways in which we can be able to respond, but the process does not end with grief and mourning. It is also about how to move forward. As Donna Haraway writes, 'mourning is about dwelling with a loss and so coming to appreciate what it means, how the world has changed, and how we must *ourselves* change and renew our relationships if we are to move forward from here'.[10]

Personal Narrative #4: Rozi

As a little girl I was told that if I worked hard enough, I could become whatever I wanted to be. Now in times of climate change, I realize this was a lie. A well-meant lie but based on little concern about the gravity of climate change for our future.

Still, part of me wants this lie to be true. I'm a half-breed: part boomer, part climate-hero. The boomer in me wants to be successful, have that house in Switzerland and live in comfort and ignorance. Whereas the climate-hero in me accuses the boomer non-stop of being out of touch with our time. 'How dare you!?' she says. Split between these two voices, I detect guilt but no shame. The aspirations of my 'boomer' are so fixed and accepted in our society that I simply cannot feel the social pressure of shame. Likewise, I have the luxury of being able to ignore any direct need to become a 24/7, full-blown climate-hero. The person I am now is somewhere between climate hero and 'boomer'. Even though my professional life is about climate change, it is a good facade for my 'boomer' side. I do not experience the transitioning of worlds just yet. I want the climate hero to win out of duty, out of responsibility, but at the same time I don't feel the existential threat to go on that journey.

Heroes of the in-between

The end of our world is also our new beginning. We are inhabiting a transformation, an in-between, the old is dying but the new hasn't been born yet. It is something of a *worldless* place. The in-between is muddy, troubled, and opaque. It does not offer a lot of ground for knowing what is what and who is who. We still have to live and experience its stories. Its lack of structure also is a creative ground: we can still shape the future, albeit not in any way we want.

The in-between pushes us to learn to think, respond, imagine, and be with ourselves and with others differently. It brings us into a transformation: an undoing and redoing of ourselves. Such a transformation is scary, unsettling, and confronting. We need courage to step into it and endure it. We need to become heroes, heroes of the in-between.

The hero of the in-between is not the old-school hero we know. The old-school hero possesses exceptional superpowers, invulnerability, independence, and a touch of vanity. With these qualities, he (usually it is a he) typically saves the world – but that world is exactly the one we have to let go of. The in-between demands heroes with different qualities.

For Campbell, scholar of mythology, a hero is not born a hero. Rather, what makes a hero is the journey. This is not any journey, though, but an *initiation* into unknown territories and transformative experiences. The journey to become a hero then is to step out of the old, confront the unknown, transform in the process, learn from the experience, and bring it back home.[11]

The heroes of the in-between, then, are not rare and exceptional people. They are ordinary persons responding to the invitation to a journey of initiation. All of us can be heroes of the in-between. It does not require special skills or gifts. We only need to answer the call and see what lies ahead of us on our journey.

If the old hero saves the world from one great threat, the new heroes live with threats. The in-between does not present us with one big problem to solve. It is about continuously learning to live with unpredictable tensions, ruptures, and moments of crisis. Rather than the power to save the world, the heroes of the in-between need to be responsive, and adaptive to new challenges in everyday life. They learn to stay, as Donna Haraway puts it, with the trouble.[12]

If the old hero is autonomous, the heroes of the in-between are entangled, relational, and co-dependent. For the heroes of the in-between, entanglement is learning to work from the fundamental overlaps between human and non-human life. Relationality for the heroes of the in-between is about recognizing that agency is distributed not to individuals but to their relations with others, human and non-human alike. The heroes of the in-between are not soloists but work together. This re-distributed agency is key to transforming our way of life. Co-dependency means that the survival of one species is the survival of the other. In the in-between, together 'we are not defending nature we are nature defending itself'.[13]

If the old hero was strong and courageous then the heroes of the in-between are vulnerable. Vulnerability comes with what Brené Brown calls 'the courage to show up when you can't predict or control the outcome'.[14] The in-between confronts us with exactly this, a dangerous and uncertain place. However, in this place, the heroes cannot shield up, because it requires openness, curiosity, and creativity so that a new way of life can be discovered. In the in-between, vulnerability becomes strength.

To become heroes of the in-between is to accept the invitation to journey beyond the end of the world. This is not an easy invitation to accept. We might have to overcome emotions of hesitation, apprehension, and refusal.

We may be misunderstood or judged by others. We may feel like we are not up to the task. Yet it is precisely embarking on this journey anyway, that makes us able to respond and to become heroes of the in-between.

Suggestions for further reading

B. Latour. *Down to Earth: Politics in the New Climatic Regime.* London: Polity Press, 2018.

R. Scranton. *Learning to Die in the Anthropocene: Reflections on the End of a Civilization.* San Francisco, CA: City Lights, 2015.

Seeds of Good Anthropocenes. https://goodanthropocenes.net/

III

Historical Consciousness

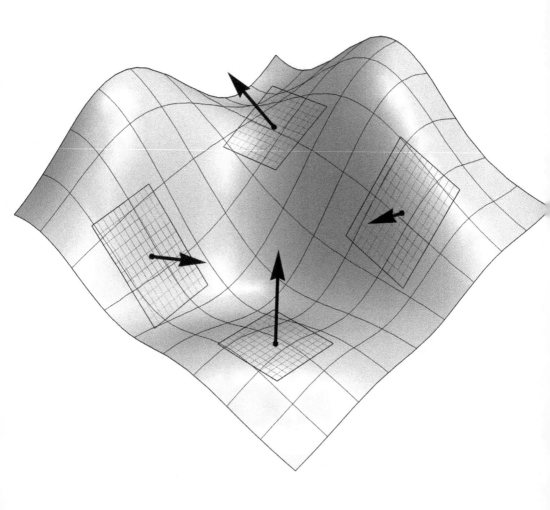

What history's most overqualified calculus student tells us about liberal arts mathematics

Viktor Blåsjö

In the spring of 1672, Gottfried Wilhelm Leibniz arrived in Paris. History remembers him as one of the foremost philosophers and mathematicians of his century. But such fame was still decades away for the baby-faced Leibniz who stepped off the coach in Paris for the first time on that chilly March evening. He had completed a doctorate in law at a remarkably young age and was sent to Paris as a junior diplomat. But Leibniz had greater ambitions in mind than the treaty negotiations he was sent to facilitate.

Paris, at that time, was an exciting place for a voracious scholar and philosopher such as Leibniz, who would later wish that he had 'twenty heads'[1] to pursue all the thinking and study he wanted to do. The Paris intelligentsia, meanwhile, were delighted to have such a passionate conversationalist enliven their dinner parties with his bold and original ideas.

Leibniz felt right at home and would have stayed for life if he could. But times were changing. French politics soured. Foreigners were driven out of the country and the once sparkling Paris intellectual milieu lost a number of its brightest stars. Leibniz had to return to Germany. For the rest of his life, he did his mathematical and scientific work in his spare time when he wasn't too busy with his day job as a courtier at the House of Hanover.

But during those beautiful few years in Paris, Leibniz formed friendships that would last a lifetime. Christiaan Huygens – the world's foremost mathematician and a key member of the Paris Academy of Sciences – had taken Leibniz under his wing. Under the tutelage of this maestro, Leibniz got obsessed with mathematics and was quickly becoming one of the field's most creative minds. His frantic notes from the Paris years show the first fledgling steps of the calculus – mistakes and all – that hundreds of thousands of students retrace in modern classrooms every year.[2]

With the expulsion of Protestants from France, Huygens withdrew to his family mansion in the Netherlands. The building – Hofwijck – is still there. If you take the train from Utrecht to The Hague you get a good look at it on your left-hand side as you approach the city. But Huygens did not retire to feed the ducks in his estate gardens. Though old and frail at this point, he kept up with the latest mathematics. This meant learning the new calculus developed by his former protégé Leibniz. The student had become the master, as the saying goes. But perhaps more intriguingly, the master had become the student.

What a treat of history this is. Reading the correspondence between Huygens and Leibniz during those years, we get to see learning in action. We get to see how the calculus is taught by its inventor, and how a sage mathematician of the highest credentials goes about learning it. We get to see the former director of scientific research at the Academy of Sciences take a seat in the front row of Calculus I, pencils sharpened and notebook in hand. It's a naked view of calculus genesis, unique in history.

Huygens proved a feisty pupil. He was not the kind of student who merely copies out the formulas and asks for help on the homework problems he got stuck on. Even proofs and demonstrations did not impress him much. What he demanded most of all is motivation. He wanted new mathematics to be thoroughly justified, not in the narrow sense of being logically correct, but in the broader sense of being a worthy human enterprise.

A fundamental concept of the calculus is the derivative. It corresponds to the speed at which things are changing, or the steepness of tangent lines. This is tangible enough, and Huygens soon understood it. But the way derivatives are calculated in calculus suggests the possibility of iterating the procedure: to take the derivative of the derivative. Calculating such 'second derivatives' is a natural thing to do from the point of view of the inner logic of the way the calculus deals with formulas and relationships. But has the calculus thereby got lost in a formalistic indulgence – a game with formulas – out of touch with concrete meanings such as speeds and slopes? Or are second derivatives actually good for something?

That is what Huygens wanted to know before bothering to study second derivatives. He writes to Leibniz:

> I still do not understand anything about ddx [that is to say, second derivatives], and I would like to know if you have encountered any important problems where they should be used, so that this gives me desire to study them.[3]

Tell me why I would want to study second derivatives, Huygens demands. Not the formal rules for working with them, and the proofs thereof, and artificial problems specifically invented for them. No, not that. Any mathematician can make up such mathematics ad infinitum. A new mathematical theory must prove itself not by solving its own internal problems, but by proving itself on a worthy, honest-to-god problem recognized in advance.

Leibniz understands this well, and replies:

> As for the *ddx*, I have often needed them; they are to the dx, as the conatus to heaviness or the centrifugal solicitations are to the speed. Bernoulli employed them for the curves of sails. And I have used them for the movement of the stars.[4]

In other words, we don't care about second derivatives because the symbolism suggested we could do derivatives once over. We care about them because they are the right way to tackle mathematically a rich range of fascinating and important phenomena. Do you want to understand the shape of a sail bowed by the wind? Do you want to describe how planets move around the sun? Then you want to understand second derivatives.

These examples were agreeable to Huygens. Not because he thought mathematics must be 'applied' – he had done plenty of *l'art pour l'art* pure mathematics himself – but because nature has excellent mathematical taste. Any mathematical theory can show off on technical pseudo-problems specifically designed to be solvable by that method. But it's one thing to chew on teething toys, and another to cut the mustard with the real deal.

There was no burning need to find the equation for the shape of a sail. But this application shows that second derivatives have nature's endorsement, as it were. That's a good letter of recommendation, and one that is not easily forged. Thus a real-world problem is a good acid test of the worth of a new idea. As Huygens explains:

> I have often considered that the curves which nature frequently presents to our view, and which she herself describes, so to speak, all possess very remarkable properties. Such curves merit, in my opinion, that one selects them for study, but not those curves newly made up solely for using the geometrical calculus upon them.[5]

Leibniz agrees: 'You are right, Sir, to not approve if one amuses oneself researching curves invented for pleasure'.[6]

If only modern calculus books lived by the same rule! Flip to the problem section at the end of any chapter in any standard calculus textbook and you will find a thousand problems 'made up solely for using the calculus upon them' – exactly what Huygens condemns. Perhaps it should give us pause for thought when both the inventor of the calculus and its most able student ever are in complete agreement that our way of writing textbooks is stupid.

Modern students may well sympathize with Huygens again when he makes a similar point regarding exponential expressions such as e to the power x:

> I must confess that the nature of that sort of supertranscendental curves, in which the unknowns enter the exponent, seems to me so obscure that I would not think about introducing them into geometry unless you could indicate some notable usefulness of them.[7]

Leibniz shows him how such expressions can solve certain problems, but Huygens is still not impressed: 'I do not see that this expression is a great help for that. I knew the curve already for a long time'.[8] The moral, once again, is: first show me what your technical thing can do, or else I have no reason to study it. And if I can do the same thing by other means then you have still failed.

So Huygens was a critical student who questioned and scrutinized everything. It is in the nature of the liberal arts to embrace and encourage little Huygenses whenever we find them in our classrooms. As another leading mathematician, Felix Klein, later put it:

> Apart from the majority of enthusiastic students there are always a few students who are not entirely satisfied, who criticize and question. These are the ones dearest to my heart. For I see in them what I consider to be the true goal of all teaching: independent thought.[9]

Yet conventional teaching runs the risk of alienating such students by neglecting precisely the kinds of questions that Huygens wanted addressed before deciding whether the new mathematics was worthy of his time. Let us make room in our educational system for the Huygenses of today, who want their passion for learning to stem from inner conviction rather than the passive acceptance of someone else's teaching.

Mathematics has been embraced as a core pillar of a liberal education since antiquity, in no small part due to the spirit of independence embodied by Huygens. In heart and soul, mathematics is still the same creative, free-thinking, curiosity-driven field that the liberal arts fell in love with all those years ago. But it's a marriage on the rocks. In many educational settings, mathematics has grown apart from liberal arts ideals, and not for nothing.

The famous inscription above the gates of Plato's Academy – 'Let no one enter here who is ignorant of geometry' – was no token curricular breadth requirement. Plato wasn't pulling any punches when he set out the central role that mathematics would play in his vision of an ideal republic. Future rulers should study advanced mathematics for ten years, between the ages of twenty and thirty, Plato demanded. No society has yet implemented this ambitious plan. Nevertheless, it is striking that a modern higher education in mathematics, from college through a PhD, corresponds quite closely to Plato's timeline.

Ten adult years is perhaps a reasonable part of a human life to spend on education and apprenticeship. That much has remained the same since Plato's time. But meanwhile the sum total of all mathematical knowledge has grown enormously. What Plato's students had ten years to learn, we would have to master in a single afternoon if we wanted to have equal time to cover all other parts of mathematics.

Yet the desire to somehow cover 'everything' remains. That is also what Plato envisioned. He didn't mean that people should take some maths classes to pick up applicable skills that would help them get ahead in tomorrow's knowledge-based economy. No, a proper mathematical education, according to Plato, encompasses all branches of mathematics and leads to 'a unified vision of their kinship'.[10] But that was easy to say back then, well before the birth of even leading Greek mathematicians such as Archimedes and Euclid, let alone the thousands of years' worth of mathematics contributed by later civilizations.

Such forces have pushed mathematics to be taught in ever more il-liberal ways. No subject is as cumulative as mathematics, and the field has had to embrace a top-down, dictatorial modus operandi to keep up with an ever-growing body of material. Instead of dealing with topics A, B, C, the mathematician extracts an abstract core X common to them all and makes this alone mathematics. The efficiency of this approach is undeniable. But what is gained in power is lost in meaning. X grew out of A, B, C, and all its motivation and purpose lies in those roots. Yet those are precisely the ties that must be severed to enjoy the gain in efficiency. Thus, paradoxically, mathematics depends in an essential way on ignoring purpose and specifically avoiding the natural way of arriving at the ideas one is trying to learn.

This is precisely what we saw above with the second derivative. The concept of the second derivative is natural in contexts such as planetary motion, falling bodies, and maximally inflated sails. But it saves a lot of time to ignore the particulars of those contexts and teach only the abstract idea. This is an intellectual betrayal of sorts, since it covers up the actual path that led to these ideas.

Teaching only the abstracted X, only second derivatives for example, is to detach mathematical ideas from the contexts that give them life and meaning. In some ways it is indeed more efficient to study uprooted wildflowers under the fluorescent lights of a lab instead of hiking through mud and rain to see it grow in its natural habitat. Yet, to quote Felix Klein again, the romantics among us still strive to make mathematics education more akin to 'delightful and instructive walks through forests, fields, and gardens [...] without digging up the most profitable plants to replant them in prepared soil according to the principles of rational agronomy'.[11]

For the training of a technocratic workforce, the efficiency gained by putting romance and questioning aside may be a worthwhile bargain. But there should be a place in the world also for those who refuse this pact with the devil, and value their independence of mind higher than any promise of power. History vindicates such rebellious souls and suggests how to teach them. A liberal arts college is the place to keep alive the reflective approach sacrificed due to efficiency demands in other programmes.

Reinstating more organic historical paths to the ideas of mathematics enriches it with meaning, critical reflection, and contextual interconnections. Pursuing these dimensions does not mean departing from core mathematics for the sake of a liberal arts perspective; rather, it means remaining true to the spirit of the very pioneers of the subject, such as Huygens and Leibniz.

Suggestions for further reading

P. Lockhart. *A Mathematician's Lament: How School Cheats Us Out of Our Most Fascinating and Imaginative Art Form.* New York, NY: Bellevue Literary Press, 2009.

J. Stillwell. *Mathematics and Its History.* New York, NY: Springer Mathematics, 2010.

S. Strogatz. *Infinite Powers: The Story of Calculus.* London: Atlantic Books, 2020.

Statistics: The art of seeking sense in numbers

Guus de Krom

In those days a decree went out from Emperor Augustus that all the world should be registered. This was the first registration and was taken while Quirinius was governor of Syria (Luke 2: 1-5).

The history of data gathering: the actors on stage

The history of data gathering goes back for centuries, as the census referred to in the Gospel of Luke testifies. This census served to obtain information on the size of the population – an important issue for any ruler. Some 1,000 years later, William the Conqueror sent his men all over England and parts of Wales to do a survey of the land, later recorded in the Domesday book. While in ancient times data collection still primarily served the interest of the ruling class, over the centuries, more and more parties became involved in gathering numerical data. The rise of trade and commerce played an important role in this, since large-scale commerce requires elaborate and accurate bookkeeping. In the seventeenth century, for example, companies like the VOC (Dutch East Indies Company), with activities spanning three continents, heavily relied on numerical data. Data is of course still gathered by states, local authorities and large companies such as Facebook and Google, but also by international bodies such as the WHO, NGOs such as Greenpeace, patient organizations, smaller companies, and individual citizens. Nowadays, almost anyone somehow takes an interest in numerical information. The historical pattern is evident: data collection has been democratized.

Coming of age of statistics as a science: development of probability theory

The recognition of statistics as an academic discipline is fairly recent. In the nineteenth century, Guy describes what, according to his contemporaries, statisticians were supposed to do – and what not. 'The statistical labourer was not to be indulged with the luxury of opinions; he was to be a patient drudge, binding up his sheaves of wheat for others to thresh out'.[1] The work was supposed to be 'humble and unintellectual'. Guy did not agree, and argued statisticians could be involved in all kinds of topics, provided they were 'rich in facts and figures'. He thought statistics should not be seen as 'a mere synonym for the collection of facts', but as a science, like others.

Guy's claim that statistics was a 'real science' could not have been made without the development of probability theory. From the seventeenth century onwards, scientists like Fermat, Pascal, and Huygens, and later Laplace, Bernoulli, and Bayes laid the mathematical framework for the computations and models used in statistics. Probability theory allowed researchers to add a crucial new property to their numbers: next to the value itself, they could now compute the likelihood to actually find that value, given certain model assumptions. Probability theory allows researchers to distinguish between values quite likely to be expected, and values that somehow stand out as *un*likely, and therefore potentially interesting. This opened the door to 'inferential statistics' – the branch of statistics that deals with the testing of hypotheses using observed data. Stigler described the development of statistical reasoning between 1700 and 1900 as a simultaneous horizontal and vertical development: 'horizontal in that the methods spread across disciplines [...] vertical in that the understanding of the role of probability advanced as the analogy of games of chance gave way to probability models for measurements, leading finally to the introduction of inverse probability and the beginnings of statistical inference'.[2]

It is interesting to note that the concepts of likelihood and error variability had actually been known for ages, but that it took centuries for these concepts to make their way into formal statistical reasoning. As an example, Stigler mentions that, ever since the Norman conquest, the Royal Mint stored daily samples of coins in a box to assess the integrity of the coinage (the 'trial of the Pyx'). The contract between the Mint and the King stated that a certain difference between the weight of a single coin and the aggregate weight of the Pyx contents was tolerated, providing historical evidence of 'an institutionalized allowance of uncertainty', hence an understanding that outcomes are prone to error variability. Stigler notes, however, that such trial procedures were not applied to other purposes, though they helped lay the foundation for a logic that was later more generally applied to data.[3]

Widening the scope: From concrete to abstract

Initially, data collection boiled down to quantifying concrete things – cattle, land area, people. The goal was a factual – possibly complete – description of the state of affairs, sometimes referred to as political arithmetic. Gradually, the aim moved from tabulations of observable plain properties to the characterization and measurement of more abstract, multidimensional phenomena. Researchers became more interested in measuring 'health', rather than just counting 'diseased people', so to say. When statisticians no longer felt bound to just observe and describe concrete properties, all kinds of phenomena became of interest. Statisticians were greatly inspired by advances in the natural sciences and believed that they, too, could discover universal laws – if only the data were analysed properly. This shift from concrete variables to more abstract phenomena constitutes a third dimension, next to the horizontal (spread across disciplines) and vertical (sophisticated development of probability models) ones described by Stigler.

Freed from self-imposed constraints to only look at readily observable phenomena, statisticians in the nineteenth century started to work on 'social physics'. One of the leading scientists in those days was Aldolphe Quetelet, a Belgian astronomer, mathematician, statistician, and sociologist (as such, a fine example of a Liberal Arts student), who introduced the idea of 'an average person'.[4] The influence of Quetelet can still be seen today in the 'body mass index', which he helped develop. Next to measures of health, researchers became interested in the idea of measuring other abstract properties, such as human intelligence. In *The Mismeasure of Man*, Stephen Jay Gould provides a fascinating historical account of attempts to define and measure human intellectual capacity.[5] The newly formed ideas of Darwin made their way into the study of humans – though heavily influenced by racist and sexist prejudice, as Gould points out.

Anything that exists can be measured – or not?

Nowadays, it seems as if people believe that anything can be measured, and subsequently statistically analysed. Consider political science. As the statistician Moore puts it: 'I find it hard to think of policy questions, at least in domestic policy, that have no statistical component. The reason is of course that reasoning about data, variation, and chance is a flexible and broadly applicable mode of thinking'.[6] Many people share Moore's belief that statistical tools are versatile – so versatile you can get any result you want ('lies, damned lies, and statistics'). Best argues that part of this negative public sentiment towards statistics is because people are naïve about the social context in which scientists always operate: 'The image is that statistics are real, much as rocks are real, and that people can gather statistics in the way that rock collectors pick up stones. After all, we think, a statistic is a number, and numbers seem solid, factual, proof that somebody must have actually counted something. But that's the point: people count'.[7] In the days of William the Conqueror, a count of cattle or number of people living in a shire was not contested as such: what constitutes cattle and people is well-defined, and while the counting could of course result in a wrong number, an outcome – so many cows, so many sheep – could be interpreted unambiguously. Erroneous or incomplete, maybe, such numbers were indeed 'real' in a naïve sense of the word. With statisticians increasingly dealing with abstract properties such as *intelligence, gross domestic product, biodiversity,* or even *Covid-19 infection rates,* rather than 'cows' or 'sheep' as in William the Conqueror's day, it becomes ever more important to realize that these concepts are reflections of abstract ideas in people's minds, and that what people have in mind is influenced by the culture they live in, and the current state of knowledge. Best writes: 'Numbers do not exist independently of people; understanding numbers requires knowing who counted what, why they bothered counting and how they went about it'.[8] Most of the phenomena we study are in fact somehow constructed by us – for some, one could even argue they exist *because we say so.* As Ambrose put it: 'Numbers are always seen through the lens of knowledge provided by a particular time and culture [...] "perverts" did not exist before the late nineteenth century'.[9] The 'perverts' category was in fact not the only or the last to be constructed on the basis of data. After the 1820s, researchers came up with all kinds of categories:

> New slots were created in which to fit and enumerate people. Even national and provincial censuses amazingly show that the categories into which people fall change every ten years. Social change creates new categories of people, but the counting is not mere report of developments. It elaborately, often philanthropically, creates new ways for people to be.[10]

The message is: there is no such thing as a context-free measurement in statistics – unless you restrict yourself to counting beans (and even then). In the words of Best: 'We need to worry about how statistics were brought into being. Who did the counting? What did they decide to count, and why? How did they go about it?'[11]

Information overload: the need to reduce complexity

When different types of information are collected, each supposed to shed light on something we find interesting, we often want to summarize the outcomes, wedging all information into one or a few numbers. Consider *human intellectual capacity*. We can think of many aspects that somehow indicate this – the ability to memorize, capacity for logical reasoning, maybe also indicators of social or emotional intelligence. Now think of a test that measures these different aspects and expresses each in a separate score. What we often want is an aggregate of all partial scores that somehow captures all information – the bottom line, if you wish. Some people believe an IQ score does that job for human intelligence. In economics, the *Gross Domestic Product* (GDP) is an aggregate number used to rank countries. Students (and, I fear, some admission boards), may see the *Grade Point Average* (GPA) as a valid indicator of academic performance. Universities are also somehow evaluated, scored, and then subsequently ranked. Even though such rankings are methodologically dubious, they are important – just because they are considered to be so.

However, confronted with some aggregate number, you should always ask: What does it tell me exactly? Consider GPA: A score of 4.0 indeed means that the student scored As on all subjects. A GPA of 3.0 is not that informative, however. It could be that the student hovers between an A and a C, or steadily performs at B level (note: one rarely hears of 'a straight B-student'). Also: what does it mean when 'the GDP has grown with 1.2%'? And what are university rankings based on? These examples illustrate that a reduction of a multitude of different numbers onto a few numbers in order to reduce the complexity is inevitably at the expense of the interpretability of such numbers. Heaping dissimilar pieces of information yields a blurred average that can no longer be mapped onto the very things we are interested in. Gould writes that Binet, a leading psychologist in the early days of intelligence testing research, was keenly aware that his proposed aggregated IQ scores had no inherent meaning: 'The scale, properly speaking, does not permit the measure of intelligence, because intellectual qualities are not superimposable, and therefore cannot be measured as linear surfaces are measured'.[12] There is a certain irony to it that while our knowledge about intelligence has accumulated after Binet, this bit of wisdom got lost in the process.

Attempts to wedge dissimilar bits of information into an aggregate number have not been limited to academic disciplines. In modern-day China, authorities and companies collect masses of data on individual citizens in real time, using state-of-the art techniques such as automatic face and speech recognition. The obtained data are worked into *social credit scores* that are used to modify an individual's behaviour by denying or allowing privileges, such as the use of public transportation and the ability to loan money. For many people, the idea of being under constant electronic surveillance – with potential immediate and far-reaching personal consequences – may sound eerily like Orwell's *Big Brother* coming into reality, though Western countries, too, have proposed or introduced far-reaching surveillance measures to prevent the spread of Covid-19 infections. Regarding China, Creemers points out there actually is a variety of systems in use, some employed by the national state, some by local authorities, others by large companies. The tools are diverse, but invariably so complicated that their output (and even ownership) is not directly transparent to the users. The yield of the systems is often unclear, and some systems have even been changed after protests of individuals.[13] In other countries, the proposed use of dedicated apps to control the spread of Covid-19 – a goal many support – has also met with severe public scepticism.

Another example of a seemingly advanced tool that does not really deliver is COMPAS, a criminal risk assessment tool (Correctional Offender Management Profiling for Alternative Sanctions), that aims at predicting the risk for recidivism. Dressel and Farid claim that COMPAS predictions are 'no more accurate or fair than predictions made by people with little or no criminal justice expertise' and that 'a simple linear predictor provided with only two features is nearly equivalent to COMPAS with its 137 features'.[14] Seeing that machine predications are equally good (or bad) as those of untrained humans, they state that 'results cast significant doubt on the entire effort of algorithmic recidivism prediction'. Among its shortcomings, one rings a familiar bell:

> Recidivism in this study, and for the purpose of evaluating COMPAS, is operationalized with re-arrest that, of course, is not a direct measure of re-offending. As a result, differences in the arrest rate of black and white defendants complicate the direct comparison of false positive and false-negative rates across race (black people, for example, are almost four times as likely as white people to be arrested for drug offenses).[15]

Indeed: who did the counting? How was it done?

The role of methods and statistics in a LAS curriculum

Today, unless you live under a rock, you are exposed to masses of information on a daily basis. Sophisticated analyses can be performed with regular laptop computers. One would think we are all able to make sense of numbers now. What really empowers us, however, is not having loads of data and fancy machines, but our ability to think critically and straight about numbers (see image). This is not the same as 'being good in maths'. Students often think methods and statistics is mathematics in disguise and perceive the usefulness or difficulty of such courses according to such preconceptions. True, statistical analyses can (only) be meaningfully interpreted because they have a sound mathematical basis, but any course in methods and statistics, certainly one at introductory level, would fall short of its purposes if it merely aimed at 'getting the maths right'. Moore says, statistics 'is a fundamental method because data, variation, and chance are omnipresent in modern life. It is an independent discipline with its own core ideas rather than, for example, a branch of mathematics'.[16] Students need to be mentally equipped to deal with data and outcomes of analyses. Formal training is needed for that, because we have no sound intuition for chance or probability – read Tversky and Kahnemann to see how we fool ourselves.[17] When our ancestors stepped off the African plains on their global tour, their brains evolved, but the ability to 'think methodologically' apparently did not yield a survival bonus. Consequently, intellectually sophisticated people do not necessarily think straight about data. This has implications for our teaching, regardless of our academic interests. Students with an interest in medicine, for example, had better be aware of the difference between *statistical significance* and its relation to the more mundane concept of *relevance*.[18] It probably goes too far to assume that all science students need to understand the technicalities of 'p-hacking' – but they should at least get the gist of why Head and colleagues took the trouble to investigate the issue and what it implies for research.[19] Ioannidis' paper 'Why most published research findings are false'[20] and Arnett's study 'The neglected 95%: Why American psychology needs to become less American'[21] should arouse the interest of many students – if only for the intriguing titles.

   In the end, we are all citizens, and since this implies we are all somehow data users, we all need to become numerate. As Rose says: 'democratic power is calculated power, calculating power and requiring citizens who calculate about power'.[22] If you are oblivious of a fundamental knowledge of methods and statistics, its premises, pitfalls, and – yes – its blessings, you do yourself a disfavour.

Suggestions for further reading

S. Blauw. *The Number Bias: How Numbers Lead and Mislead Us.* London: Hodder & Stoughton, 2020.

H. Rosling, O. Rosling, A. Rosling Rönnlund. *Factfulness: Ten Reasons We're Wrong About the World – and Why Things Are Better Than You Think.* New York, NY: Flatiron Books, 2018.

D. Salsburg. *The Lady Tasting Tea: How Statistics Revolutionized Science in the Twentieth Century.* New York, NY: Henry Holt and Company, 2002.

The canon of the Netherlands revisited

James Kennedy

Introduction

A canon possesses authority. A canon by definition sets the standard for what is excellent, for what is authoritative, for what is authentic. Think of the Bible – any book in it has more authority than any book not included in it, at least in the Church. Shakespeare's canon only includes works clearly attributable to the great master. Literary canons only include works judged to be of the highest quality and having the most enduring value.

Any canon, at the same time, generates controversy. There are always people who object to what's in a canon; there is always some work or some author demanding inclusion – or exclusion. Because canons attempt moreover to separate 'the best from the rest' they raise suspicion. Who decides what is 'best'? Don't canons wrongly privilege a particular theological, political, or aesthetic outlook – and with it, a particular class, religion, race, or gender? Don't they attempt to fix and freeze fluid complexities? Shouldn't we then dispense with canons altogether?

The introduction of the Canon of the Netherlands in 2006, then, inevitably sparked controversy.[1] It was a canon of history, not of books or other cultural works. Now, rather unconventionally, the government wanted a 'canon' that selected what was important for Dutch school children to know about Dutch history – events, persons, places, cultural artefacts – anything, really. I had no role in creating that canon. But I was asked in 2019 by the Minister of Education to revise the 2006 version with a commission of my choice. Although I had my doubts about the Canon of the Netherlands, I accepted the task as a worthy challenge, and in June 2020 we offered a canon with significant revisions.[2]

In revising the canon, I came more than ever to believe in the value of a canon. It is good that a society – or more particularly in this case – school children have a sense of what is most important to know about the country in which they are raised. But having a canon only has value if a) it promotes discussions about what is important; and b) its content shifts with time, so that it can optimally serve those discussions. The best kind of canon is the kind that encourages good debate.

In this article, I will outline the creation and, above all, the revision of the Canon of the Netherlands. In doing so, I will respond to critics who primarily think a canon is, happily or unhappily, an immutable measuring stick that is imposed upon society and schools.

Creation of the canon of the Netherlands and its reception

Most ministries of education do not call these essential elements in a history curriculum a 'canon' – they may be more prosaically known as 'history standards', as they were in the United States. Only two countries as far as we know have sought to call this historical corpus a national 'canon'. The Dutch were the first in recent times to do so, in 2006, and the Danes followed several years later. The Flemings are now creating their own canon, based on the Dutch model.[3] In the Netherlands, it was fifty 'windows' into the Dutch past, each window highlighting something about the Dutch past, from the first Dutch text to Vincent van Gogh and the Second World War. The Danes made do with only 29 seminal points, ranging from Danish events like the abolition of servitude in 1788 to the Treaty of Maastricht in 1992, which established the European Union.[4]

The canons of both countries were created because of the desire of politicians to cement the importance of national history in an uncertain and increasingly globalized world, and to impart this importance to future generations. The very use of the word 'canon' was intended to emphasize this intent. It may be questioned whether the Dutch Ministry of Education ever really got what they wanted. The first commission under the Utrecht professor Frits van Oostrom came up with an almost playful canon. It disclaimed a role in promoting Dutch national identity. Its status in the Dutch educational firmament was to be 'inspirational' and not the object of rote learning. Initially it was not even closely tied to the history curriculum but – with an eye to the pre-disciplinary education of primary schools – to be used by teachers from a wide variety of subjects as they see the opportunity. Later it was tied to history education's 'core aims', but still only as 'inspiration'.[5]

When the Canon of the Netherlands first appeared, it was largely sceptically received or opposed by university-based scholars and to some extent by educationists. They distrusted the efforts of national governments (and not only in the Netherlands) to reassert the importance of national history and, through it, national identity and community. A privileged 'canon' of national history, however variously constructed, was intended to inculcate students with a stronger sense of belonging to the same national community and was intended to counter 'multiculturalism'.[6]

Various arguments were heard along these lines. A canon is a form of statist indoctrination, focused on ends like cultivating citizenship rather than historical understanding. Canons are by nature fixed and closed, not all reflecting the open and fluid nature of the past. Alternately, they are too national or at least too European. In practice, a canon is too rote, unresponsive to the specific learning contexts of each classroom.

In many ways the particular construction of the Dutch canon in the years 2005-2006 was designed to parry such critiques. It was created by an independent commission not subservient to government *Diktat*. In emphasizing its purpose as an educational tool, the commission sought to distance itself from being used as a normative cultural canon for the nation, even as it recognized that it might be used as such. Educationally, its function, as noted above, was simply to be a source of 'inspiration' for teachers; its diffusion of Dutch history into fifty discrete windows or stories made it ill-suited to generate a coercive 'master narrative' of the nation.

Revision of the canon and its reception

By the time the government commissioned revamping the canon in 2019, the cultural climate had changed. The concern for strengthening national identity had not disappeared, but the emphasis lay more on the social participation and inclusion of all segments of society. The minister of education asked that we look at the representational aspects of the canon, specifically naming women. The new commission was charged with making adjustments to the existing canon rather than making any radical changes.

In the end, though, the changes it made were significant enough. The commission simplified the texts, which initially had not been written with accessibility as the chief concern, and updated the content of most windows. It connected the canon to history education through the further development of thematic lines. It cross-linked international developments more in the window texts. It sought to ensure that all parts of the canon made room for multiple voices and perspectives. Most manifestly it replaced ten windows with new themes or persons, and recast four more in a different perspective. The new emphasis on looking to widen the perspective to include more women, minorities, and other stories from the margins was the general guide for making the changes, even as the commission sought to create a well-balanced canon of multiple perspectives.

The public and particularly teachers generally seemed to approve of the changes made. Sceptics from academe and from the teaching community thought the revisions were better than they had expected. Public approval was not necessarily informed about the details of the changes. Much of the discussion was about the choices of specific windows; in the wake of the Black Lives Matter debate the choice for the activist Anton de Kom seemed to come at the right time; the 'replacement' of postwar prime minister Willem Drees in favour of Marga Klompé – a chief architect of the Dutch welfare state – was more controversial. Historian Rimko van der Maar twittered that he thought Drees 'a hundred times' more important than Klompé. In addition to these broad questions some specialists asked us to revise what they regarded as mistakes in the text – which we did if we thought it merited.

More interesting was a more architectonic critique of the revised canon. The critics of 2020 had to a very large extent a different set of concerns. One critic maintained that we were still imposing a statist pedagogy, but now promoting multiculturalism rather than national identity.[7] Others saw the inclusion of more women and minorities in the canon as having no historical value but only a therapeutic one so that some groups could recognize themselves in the past. One retired historian objected chiefly to our overly critical view of the Netherlands' slavery past; a rider to what he regarded as our feel-good inclusion of inconsequential women and of popular culture, namely sports.[8]

Reviewing objections to the canon

I have grouped three sets of objections to both canons that I have combined to make a compact series of observations. A fourth line of critique – on the canon's relation to the classroom – I am omitting as less interesting in the context of this article.

The canon as state pedagogy

Of all the objections to the Dutch canon, I think this one is most problematic. In the first place, it was independent commissions who were charged with making the canon, with the government having virtually no say in what was actually contained in the canon. The choices made were then hardly the result of a government dictate. Even more important, the canon's function as inspiration for teaching leaves a great deal of latitude for teachers to teach the canon as they see fit. That is how it should be. The canon is meant to be used critically, in which the very choices made by the commission can be questioned, and where multiple perspectives are encouraged. This dialogical approach strongly favoured by both commissions is a far cry from a government-imposed system of turning instructors into state functionaries who drill pupils to think in a particular doctrinaire way. My guess is that this particular critique overestimates the capacity of canons to shape state ends. In a highly pluralistic society like the Netherlands, any canon simply does not have the power to do so, also because schools are diverse and the canon is differently applied at each of them.

Even if I doubt that the state will be able to use a canon to indoctrinate its students, the question might still be raised if citizenship formation as an aim in history education is something to be avoided, as historians have maintained. I think, however, that it is misplaced that study of the past is entirely divorced from moral reflection. The purpose of a historical canon is to make alive the past to students in primary and secondary education; and stirring their interests in the past is about them seeing other worlds where our own moral frameworks did not exist. But to divorce the historical and moral imagination from each other makes it more difficult for students to make connections with past and present – also an important aim of history education. Our canon window on child labour laws tells something about what it once was like to be a child, but it also opens questions up about what is still going on in the world, and how the past can shape how humans might now respond.

The canon as an ideological tool

Canons are often created to serve political or social ends. The first one clearly in its political context was to shape Dutch national identity. The commission did not deny in its report of 2006 that the canon might serve legitimate goals of stimulating a sense of national identity, but that was not its own primary aim, which was to stimulate historical awareness and knowledge through education. In doing so, it effectively sought to depoliticize and de-ideologize the canon. As noted above, the specific construction of the Canon of the Netherlands avoided making large narrative claims about Dutch history, preferring the smaller, discrete stories of 50 windows. Since 2006, Frits van Oostrom has defended the canon as being educational rather than identitarian in aim, thus seeking to defang it of any political potency.[9]

Giving the durability of this particular canon this effort may be considered successful. The fifty windows were also diffuse enough not to privilege too much one vision of history over another. In contrast to what might be expected from a national canon, it was not a 'whitewashing' of Dutch history but included some of the country's more painful episodes, such as slavery, colonization, and the Holocaust, alongside the highlights of Dutch culture.

The question is whether the second commission in its seeking balance was in its own way ideologically motivated. The revised canon is more 'social' than the first one – less focused on high culture and more interested in the trajectories of social and political change, a bit more focused on the 'peripheral' regions, marginalized figures, and particularly women, because it held that these too are important for telling crucial stories about the Dutch past. This has generated the critique that the canon now includes historically less important people. They were misguidedly selected to assign importance to women or persons of colour which they did not possess in the past, and to facilitate wrongheadedly the ability of female or minority students to identify with historical personages like themselves.

I would dispute the assertion that the commission chose anyone who was less important. The point is that the canon be able to tell a wide range of different, compelling stories about the Dutch past, so that the Dutch historical experience can be shared in its richness and complexity. That we should replace Charles V, ruler of a world empire, by his grandmother Mary of Burgundy, who died at 25, was seen as choosing someone less important over someone more important. We held however that some important elements were made visible in Mary that justified removing one famous man: not only the gendered nature of political power, but also the story of how the Burgundians – absent in the old canon – were superseded by the Habsburgs, and how Mary's brief reign set the ideological grounds for the later revolt against Spain.[10]

I do not doubt that the choices for some of our windows will allow some students to more strongly identify with them. The inclusion of the anticolonial Anton de Kom of Suriname is often cited in this context. It is fine if students do so, but that was not the primary point in making our selection. Many if not most of the choices we made do not so easily lend themselves to 'identity politics' in whatever sense and our emphasis on multiple perspectives even less so. Most crucially, we have always considered the canon as a total concept, in which the parts are seen in relation to each other. Users, in our view, should not avoid windows further removed from their own perspectives or identities but rather develop curiosity about them.

In general, a canon that is more inclusive of different voices is a better way to reveal the past, and to initiate discussion about it. If done properly, this should not mean skewing interpretations of the past in a particular ideological direction but lead to a balanced canon where the range of stories is as broad as possible.

Can or should a canon be 'living'?

Much of the initial scepticism towards the canon was that all canons were fixed, unmovable. That is also precisely what some of its proponents also desired of the Canon of the Netherlands. The first commission however was clear in that it intended to function as a 'living canon', to be used in fluid ways and to be periodically revised. In the first instance they even proposed a revision once every five years – much shorter than either the fourteen years it took or the ten years the recent commission advises. I think that a periodically revised canon is good and necessary so as to meet changing needs or incorporate new insights. In some cases this makes obvious sense. One of the windows featured the Groningen Gas Fields. We now see more clearly the environmental and social problems associated with those fields than was the case in 2006. And historical research and debate shifts over time the focus we give to the past.

The more interesting question might be whether understanding the canon in this way so undermines the notion of a canon being a dependable 'standard' that it might as well not be called a canon at all. That is what has been suggested, and it is a serious point.[11] My own view is that we can call this a canon because it still makes strong claims about what Dutch students should know about Dutch history. The choices made were not arbitrary but well considered, meant indeed to identify the major themes in Dutch history and to confirm a set of windows that represented the breadth of those themes. The precedent has been set for significant change the next time a new commission is asked (if it is asked) to revise the canon. But that is not the only outcome. All or most choices made by us, or the former commission, may be maintained by commissions in the future. In that case, the further reifying of our choices is a classic process of canonization, in which the most essential is distinguished from what is less essential. I have difficulty imagining, for instance, that Anne Frank will soon be removed from the Canon of the Netherlands. This possibility is a reminder than canons do not arise *ex nihilo* but are themselves subject to longer-term, sometimes creative processes of canonization.

Conclusion

The Canon of the Netherlands in my judgment has been a boon to boosting interest and reflection on the national past. It has done that through powering the public imagination and having the proven (primary education) and potential (secondary) capacity to help Dutch students consider its past. It will have the continued capacity to do so as long as it proves itself to be open to multiple perspectives and to evolving debates and shifting interests in the history of the Netherlands. No historical canon should be – can be – a measure of permanence that establishes for all time what is worth knowing about the past. It is enough that it serve as a standard for our own time.

Suggestions for further reading

Canon Revision Commission 2020. *The Netherlands in a Nutshell: Highlights from Dutch History and Culture.* Amsterdam: Amsterdam University Press, 2020.

M. Grever, S. Stuurman. *Beyond the Canon. History for the Twenty-First Century.* Basingstoke: Palgrave, 2007.

L. Symcox, A. Wilschut. *National History Standards: The Problem of the Canon and the Future of Teaching History.* Charlotte, NC: Information Age Publishing, 2009.

Love thyself: An abridged history of Western portraiture

Tijana Žakula

With the inception and advent of social media, portraiture and its sub-category self-portraiture have become ubiquitous. Little wonder that the questions of how to strike an advantageous pose, choose a suitable backdrop, convey one's fabulous and enviable lifestyle that everyone wants to emulate have come to occupy an important place in everyday life of avid Facebookers and Instagrammers. So much so that successful attempts at self-representation have been labelled as Instagrammable portraits or selfies. In spite of their evident popularity, not everyone approves of this practice and what it entails. How many times have you heard people criticize pouty-lipped duck-faced pictures, too gaudy outfits, a tad too much flesh shown, or too much faith in filters? One easily forgets, however, that this development and the criticism thereof are by no means a novelty. In terms of self-love and self-promotion, attitudes towards fashion and idealization, albeit captured in different media, not much has changed since classical antiquity, and can be traced back to the outset of the Western concept of portraiture. Its difficult pedigree ensured portraiture's low position on the hierarchical ladder of artistic genres and encouraged art theorists to sneer at painters who specialized in this branch. In their view portraitists not only painted what they saw without keeping an eye on invention, but also, and perhaps even more dubiously, they followed their sitters' caprices, oftentimes beautified their appearance, and hence preferred money to art, i.e. were lacking in moral fibre.

Gods, Lovers, and Emperors

The very first image recorded in Western sources was a portrait. And not just any portrait, but the one of the handsome youth Narcissus, who, having seen his reflection in the water, became so tantalized by his own looks as to be unable to leave the place of self-worship until he metamorphosed into a flower. This meant that the category of portraiture, perhaps most accurately explained as an image of a person which is made in order to be recognizable, has been stained with a label of self-love, and hence vanity

– the mother of all sins, ever since Ovid (43 BC-17/18 AD) related the myth in his Metamorphoses.[1]

The first known art theoretician, Leon Battista Alberti (1404-1472), whose aim was to invest the art of painting with the dignity of a liberal art, also had something to say about one's likeness captured in visual arts. In perfectly sensible terms he explained that it contained a divine force which not only made absent men present, but moreover made dead men seem almost alive. By so claiming, Alberti tapped into some important uses of portraits, for, indeed, images have always been used as substitutes for real people in a plethora of ways. Sorcerers, for instance, would cast their spells through likenesses, while enraged people would often vent their anger by destroying the images of their tyrants. In all likelihood, though, the principal purpose of portraiture was always to capture the absent sitter's features whether he or she was dead or alive. To illustrate this Alberti recalled the story related by Plutarch that Cassander, one of Alexander the Great's captains, having been startled by the lifelikeness of the image of his king, trembled upon seeing his portrait.[2]

Alberti's vivid account not only underscored the crucial aspect of portraiture, but also raised a very important question of who was entitled to be portrayed. Pliny the Elder (23-79 CE) had already written about the art of portraiture, and he did so through stories referring only and exclusively to rulers. This is perfectly understandable if one takes into account that only emperors could have their features depicted on the obverse of coins. The first one to recognize the potential of this medium was Julius Caesar (100-44 BCE), who made himself omnipresent and always visible to his subjects by having his profile minted on coins. Apart from this, Pliny also approved of a practice established by the kings of Alexandria and Pergamum, who started portrait galleries of celebrated writers, poets, and philosophers to deck out their libraries.[3] This tradition was re-established in Renaissance Italy, albeit in a slightly altered form. Around 1370, in the *Palazzo del Capitano* in Padua, a hall was frescoed with portraits of the city's most illustrious men, launching a fashion that was to remain in vogue in the centuries to follow.[4] The only portraits of common people that were allowed were those of lovers who were far away from one another. According to Pliny, the daughter of the Sycion potter Butades traced her lover's features on the wall before he parted from her, and in this way preserved her beloved's likeness to warm her heart during his absence.[5]

The sixteenth-century Portuguese painter and humanist, Francisco de Holanda, agreed to a certain degree with his influential predecessors' views, and noted that only the distinguished deserved to have their portraits painted. In his 1549 treatise, De Holanda sanctioned that people who merited the praise of being immortalized in the visual arts were illustrious princes, kings and emperors, princesses and queens of virtue and wisdom, men famous in arms, art, and letters, or of a singular liberality and virtue, and nobody else at all.[6] This had long-lasting consequences on how likenesses of people who did not belong to any of the aforementioned categories were perceived by art theoreticians and connoisseurs.

These ideas not only seem to have come to life, but they were also radicalized through the theory and practice of the French Royal Academy of painting and sculpture (*Académie Royale de Peinture et de Sculpture*). Its founding document emphasized that the portrayal and glorification of the king, at the time Louis XIV, was the main raison d'être of this institution. On the festive occasion of its establishment, on 20 January 1648, the painters' spokesman, Martin de Charmois, presented the artists to the *Roi Soleil* as new Apelleses ready to immortalize his majesty's noble features in the most comely manner. Hyacinth Rigaud's portrait of Louis XIV (1701, now in the Louvre) is a good example of how this was supposed to be done. Even though the figure looks credible and sufficiently lifelike at first glance, the chief purpose of this portrait was not to show what the 63-old ruler actually looked like: more than anything, Rigaud's image was intended as an expression of his absolute power. In this picture Louis XIV is accessorized with the appropriate regalia: the purple curtain that reveals his royal majesty flaunting a pair of shapely ballet-dancer legs, the royal sceptre in his right hand, the coronation robes embroidered with the fleur-de-lis and lined with ermine, the sword at his left hip, as well as the crown casually placed on the cushion. In this manner, the ruler's likeness, surrounded by his attributes of grandeur and glory, transcended mere representation, and through the mixture of imitation and invention became the equivalent of an allegory.[7]

Quod licet Iovi, non licet bovi revisited: tailors, butchers, and the nouveau riche

As one may expect, practice defied theory and proved that not only people of merit had their likenesses painted. Especially as of the sixteenth century, when the output of portraits rose dramatically. They were a mark of distinction, indicating financial success, and bore witness to the sitter's social aspirations. Such pictures became available to a broader audience of people who had gained wealth and now dared to imitate and emulate the nobility through the agency of portraiture.

Little wonder that writers on art strongly disapproved of this practice. In July 1545, in a letter to sculptor and medallist, Leone Leoni, the famous

Renaissance man of letters, Pietro Aretino, condemned the century in which tailors and butchers could have their likenesses painted, thus indicating that Moroni's *Tailor* of around 1560, now in the National Gallery in London, might not be an isolated example.[8] As far as for the portrait itself, even though the subject's colourful attire clearly indicates that he was not an aristocrat but a middle-class professional, the choice of accessories suggests that he was held in high esteem. His leather belt with a buckle from which a sword could be hung, a ring with a stone on his little finger, as well as the fact that he could commission and pay for a portrait like this, identify him as a respected craftsman – a sort of sixteenth-century Gianni Versace perhaps.

Giovanni Paolo Lomazzo, an artist and theoretician from Milan, was of the same opinion as his fellow-countryman Aretino. In his Treatise on Painting (*Trattato dell'arte della pittura*, 1584), Lomazzo wrote that portraiture became vulgarized through the images of people who did not merit the honour of being portrayed, and that this tendency would ultimately degrade the entire genre. For Lomazzo an improper model meant an insignificant image.[9] Portraiture was thus in danger of being deprived of its dignity, and so was the painter, if he went along with the caprices of his ignorant sitter.

Had any of these authors been in a position to witness the unprecedented flowering of portraiture in seventeenth-century Netherlands, they would probably have been completely stunned by the abundance of portraits, and even more so by the pedigrees of the sitters whose imperfections were meticulously revealed in accurate likenesses. This love of verisimilitude sprang from the Earlier Netherlandish tradition of creating lifelike images, in which the novel medium of oil paint allowed for vivid representations, showing the sitter in three-quarter or full-face view.

From the art-theoretical point of view it was in the Netherlands where the glorious history of portraiture was once and for all challenged and deprecated by the vulgar. Gerard de Lairesse (1640-1711), the artist and theoretician from Amsterdam, was sneering at the rich, who, spurred by pride, demanded that their likenesses be painted so that their children could boast of their progenitors' deeds regardless of how insignificant or ludicrous these actions may have been.[10] One ought to bear in mind that having one's ancestors immortalized in pictures underlined their genealogical continuity, which was hitherto restricted to hereditary nobility. This practice was thus a visible testimony to the wish of Dutch moneyed burghers to emulate people of title.

An equally reproachable custom was immortalizing the new generation. In this respect De Lairesse informs us that moneyed elite would have their children portrayed for the first time at six months old, the second time when they would reach the age of ten, and ultimately at 25 years of age.[11]

Much like his illustrious predecessor Pliny the Elder, De Lairesse allowed portraits of common people only if a lover was absent from his mistress. They were encouraged to send the likenesses to one another so as to be able to cherish and increase their love of one another.[12]

Practice defied theory, however, and many artists pursued this avenue, for portraiture was the only specialization that was made on commission in the Northern Netherlands, and meant secure bread and butter for the painter. Obeying the quirks of the entitled sitter was potentially dangerous though and could jeopardise the painter's reputation. De Lairesse thus deemed it perfectly reasonable to warn the portraitist of the nuisance of having to cater to the caprices of his predominantly nouveau-riche clientele, and advise him how to go about it. The painter was well advised to insist on having his say on the point of decorum, and for his own reputation's sake introduce only suitable props into the portrait setting. These would ideally enrich a portrait and make it look more noble.[13]

There were two ways of introducing props into the portrait. One consisted of adding appropriate accessories to the background, while the other entailed a complete makeover that historicized both sitter and setting. The former meant that a philosopher would be accompanied by a celestial globe, while an orator would be depicted with a statue of Mercury. The latter, on the other hand, the so-called *portrait historié*, by being staged in a historical or mythological setting also encompassed the concept of timelessness.[14]

Fashion has indeed remained one of the archenemies of portraiture. Costly gowns were of great importance to many sitters, for they demonstrated affluence and social status. Some of them would spend a small fortune on the latest fashion in order to show off their wealth and reputation. Today's hottest buy could look completely ridiculous the next morning, however, for even the heftiest price could not guarantee the timelessness of a dress.[15]

Dressed for all times and seasons: from Van Dyck to David

There were three principal approaches to harnessing the fickle mood of fashion in portraiture. Apart from the aforementioned *portrait historié* in which the sitter would be represented as a mythological character, he or she could also opt for being represented in the nude, or cloaked in a shapeless dress.

Even though it was usual in Greece to represent heroes stark naked because the human body was considered supremely beautiful, over time this mode of representation became reserved for the bravest only. Especially if the likeness remained unidealized. In 1776, in the name of artistic freedom, Voltaire allowed the sculptor Jean-Baptiste Pigalle to represent him entirely naked except for the flowing drapery that crossed the ageing philosopher's left shoulder and covered his loins. Such a portrayal was unprecedented in the early modern period. It not only was the first sculpted portrait of a living author, which was hitherto a royal prerogative, but it also showed the deteriorating body of the great thinker in all his decrepit glory. Once heavily criticized, the uncompromising naturalism was later perceived as the ultimate victory of mind over matter.

If the sitter preferred to be covered, however, the Flemish painter Anthony van Dyck had forged a partially fictional loosely draped costume, which was understood as a classicising gown that was fully equipped to defeat the robes of ever-changing fashion. During his stay in England from 1632 until his death in 1641, Van Dyck painted a number of portraits of people of the highest rank, who were all clad in comely, seemingly timeless gear: men were wearing Japanese gowns, i.e. kimonos, while ladies would usually resort to wearing a draped piece of clothing over their gowns (see image). Kimonos were first brought to the Netherlands by officers of the Dutch East India Company in 1609, to later take the European fashion scene by storm, after tailors in England started making similar robes. There were two reasons behind their popularity. The so-called Japanese gown was very costly and fashionable, while its exotic shapelessness and loose folds evoked the simplicity of the Roman toga and gave the sitter a sophisticated look with an air of timelessness.[16]

The only remaining aspect of the portrait that would still easily get dated was the sitter's hairdo. The first painter to have successfully solved this problem was Jacques-Louis David. His Madame Récamier, painted in 1800, came to epitomize the timelessness in portraiture. In this true classic the Parisian socialite Juliette Récamier is reclining on an elegant Directoire sofa in front of an almost monochrome backdrop. She is barefoot, wearing a shapeless and yet exquisite Empire dress, while the tresses of her luscious curly hair are nonchalantly held in place by a simple black ribbon.

David's style was successfully emulated, and perhaps even surpassed by his most talented student Jean-Baptiste-Dominique Ingres. However, Ingres's portraits also marked the comeback of high fashion, that was celebrated in numerous meticulously recorded gowns worn by the rich and powerful.[17] Quite telling in this respect is what the critic Théophile Gautier wrote of Ingres's portrait of Betty de Rothschild in 1842:

> It would be difficult to make a personality and a social position bet-
> ter understood [than Ingres did] with the choice of the pose and the

arrangement of the costume. The artist had to paint a woman of the world; the world that bathes in an atmosphere of gold; he knew how to be opulent without being ostentatious, and he corrected the sparkle of the diamonds with the flash of intelligence and wit.[18]

Even though Ingres, unlike his predecessors, was very much in control of the choice of sitters, he still fostered the same love-hate relationship with the artistic genre recognizable in earlier musings on art. In his letter of 26 June 1842, he intimated to his friend Jean François Gilbert: 'And Tuesday, I have a definitive sitting with Mme De Rothschild, which came at the price of a dozen puerile and sincere letters. Long live portraits! May God damn them!'[19]

Selfies, pockmarks, crooked noses, and filters

And indeed, even though portrait painting remained lucrative, the portraitist was always expected to steer between the Scylla of naturalism and the Charybdis of idealization. If we encounter in the portrait of Mme Rothschild a joint a bit too thick at the left wrist, it is apparently because 'M. Ingres would never permit himself to omit a defect that he had before his eyes' as the aforementioned art critic Gautier pertinently, albeit naively suggested. Slavishly following nature was never part of the portraitist's project. Had Ingres indeed had a chance of representing Betty de Rothschild's wrist in a more elegant way without being called a fraud, he would have certainly done so.

Madame de Rothschild's graceless wrist may well have fallen into the category of the so-called natural deformities that were 'commendable', for they helped the likeness. On the other hand, the accidental ones, such as a blind eye, a wound, a wen, a mole, pits of small-pox, or acne, were not to be represented. These flaws would look even worse in the picture, and, to prevent such a disaster, the painter's duty was to manipulate reality mildly and represent life from the handsome side. It is therefore not surprising that one of the portraitist's main tasks was to instruct his sitter how to strike a pose revealing only the best parts of his or her physical appearance. By doing so the portraitist's moral conduct and reputation were to remain immaculate. Although the artist was not telling the entire truth, neither was he or she lying.[20]

With the invention of the daguerreotype in 1839, lifelike portraiture moved to the photographer's studio. Slowly yet surely the realm of painting and sculpture became an arena for endless experimentation with form, and only the bravest, or those who had a vested interest in promoting art of this new breed of artists, would dare sit for Picasso and his colleagues, who would more often than not make their sitters unrecognizable even to their nearest and dearest.

Ladies and gentlemen of note wearing their best frock were flocking to photographic studios to have their likenesses immortalized in this new medium. Depending on the occasion a photographer in charge would decorate his studio with appropriate props that would additionally explain the occupation of the sitter, or underline their wealth and importance.

It was in 1839 that the very first documented photographic self-portrait was made. It was taken by the photographer Robert Cornelius in his studio in Philadelphia. Be it because Cornelius was trying to adjust his daguerreotype, or he was keen to analyse his likeness, or make an artistic statement, it is safe to say this photo pre-empted the world's latest phenomenon in the domain of portraiture – the (in)famous selfie. Apart from this, Cornelius's work also abutted on the centuries-long tradition of self-portraiture. Many painters practised this sub-genre for various reasons: to have their likeness exhibited at the Medici's gallery, to make their looks known to their fans, or because their own features were always there when the lack of financial means would not allow them to hire a model.[21]

And much like old masters, we keep experimenting: with facial expressions, with backdrops, with fashion. Depending on whether one wants to make him or herself recognizable to their former classmates, or promote their good looks. What we may learn from history, though, is that the best portrait might well be the one that represents what we look like in our daily doings, wearing what we normally would. Remaining true to oneself is perhaps what constitutes beauty in real life. With or without make-up, or most expensive frock. Being true to oneself is quite possibly timeless too.

Suggestions for further reading

L. Campbell. *Renaissance portraits: European portrait painting in the 14th, 15th and 16th centuries.* New Haven, CT: Yale University Press, 1990.

E.S. Gordenker. *Anthony van Dyck (1599-1641) and the representation of dress in seventeenth-century portraiture.* Turnhout: Brepols, 2001.

É. Pommier. *Théories du portrait: de la Renaissance aux Lumières.* Paris: Gallimard, 1998.

T. Žakula. *Reforming Dutch Art: Gerard de Lairesse on Beauty, Morals and Class.* Amsterdam: Stichting voor Nederlandse Kunsthistorische Publicaties, 2015.

Two Monuments

Mary Bouquet

Introduction

Monuments are a well-established means of relating to and materializing the past, in the present, with an eye to the future. While some see monuments as being primarily concerned with memory,[1] others define them as a sub-category of public art – art in the public domain aimed at inspiring the community.[2] Monuments assume widely different shapes and forms across the centuries and have correspondingly diverse meanings: from prehistoric monumental complexes such as Stonehenge, to edifices making civic virtues visible to an urban community – such as the Parthenon in Athens. Statues installed in the far corners of the Roman Empire reminded populations there of the glory of Rome. Medieval cathedrals inspired communities as monumental works of art. Renaissance artists drew inspiration from the ruins of Antiquity to create new artworks. The Monument to the Fire of London was designed to celebrate the rebuilding of the city after its devastation in 1666. Historical monuments became powerful means of creating national memory after the French revolution broke with the Ancien Régime: the Musée des Monuments français began as an effort to assemble monumental tombs threatened by the closure of religious institutions in Paris. The Statue of Liberty was a gift by the people of France to the United States in 1886 commemorating the friendship of the two nations. And in the twentieth century, artworks such as Zadkine's 1951 Destroyed City, a figure whose head, arms, and legs explode in different directions, embodied the havoc wrought on Rotterdam by the bombing in 1940. The statue has become a monument expressing the hope of rebuilding in the post-1945 era. All these examples, and many more, suggest that there is a strong relationship between monuments and art, as well as mixtures of original intentionality and historical consciousness encroaching later on. Even if an artwork or a site does not start out as a monument, it may end up becoming one – through the passage of time and the stirring of memory that brings the past to life as part of the present – and the future.[3]

This essay will focus on the dynamic traffic between a public artwork and a monument through a case study of a specific historical episode in Groningen, the Netherlands. The discovery of natural gas in Groningen in 1959 was commemorated in 2009 by the unveiling of a public artwork celebrating 50 years of prosperity. The slow underswell of criticism that led to the formation of an activist group in 1963 and a whistle-blower who made the link between gas extraction and the increasing frequency of earthquakes was initially met with official denial. Gradual acknowledgement and, ultimately, recognition of the causal link, led to the placing of a monument to memorialize the other side of the gas story in 2019.

Rather than theorizing the distinction between public art and monument, I was prompted to try to imagine how a conversation between these two would go. I took my cue from the commissioners of the 2019 monument who explicitly refer to it as *answering* the pre-existing artwork.[4] This dialogue drew upon the forms, materials, accessibilities, and aesthetics of the artworks, as I observed them; as those directly involved with them in different ways discussed them with me; and through the secondary sources I consulted. It also led me to bring in a third protagonist. Conceptualizing artworks as persons has a long history in Anthropology,[5] and resonates with the more recent material turn which takes seriously the agency of images and things in social life.[6] Before the monumental conversation, I'll first sketch the historical context in which the two monuments came into being.

The two monuments

On 28 May 2019, a hollow, 2 x 4 m rust-coloured monolith made of weathering steel, with a vertical crack from its base to its summit, was unveiled beside the A7 in Groningen. This Other Monument counterbalances an existing artwork, the Slochter Molecule, 6.7 km away on the same road – north of Hoogezand. The accompanying image unites the two artworks in an imaginary landscape. The Slochter Molecule gambols above the horizon in the middle distance; while the Other Monument occupies the centre-right foreground, a rectangular shape firmly anchored in the ground, the upper mouth of its crack silhouetted against the sky. The vanishing point of the motorway helps propel the Slochter Molecule into a preterite position. In reality, the monuments are experienced sequentially and at speed.

Natural gas was discovered in a field near Kolham in 1959 and became a major source of energy and national revenue over the next 50 years. The gas field contained 2800 billion m^3 gas in 1959; by 2012, 780 billion m^3 remained.[7] The significance of gas revenue for national prosperity was publicly acknowledged by the Dutch Oil Company (hereafter NAM) in commissioning an artwork to commemorate the 50th anniversary. The Slochter Molecule (8 x 8 x 6 metre), by artist Marc Ruygrok, was presented to the municipality of

Slochteren and installed in 2009 on the central reservation of the motorway south of the location where gas was first discovered. The Slochter Molecule looks lightweight – like a blown-up, methane molecule – although it is made of polyester reinforced with fibreglass, resting on a steel construction. This magnification experienced at speed results in a feeling of exhilaration.

Yet for many who inhabit this landscape, the Slochter Molecule was a vexing daily reminder of the long-term damage to houses and property, to livelihoods, and mental health, resulting from the earthquakes. Vandals, who did manage to access the artwork, daubed it with paint. So, while some whizzed past, others infiltrated and vandalized; and yet others were provoked to respond differently.

The Other Monument 'calls attention to the earthquakes and their disastrous consequences'.[8] When gas is extracted from a porous layer of sandstone 3 km beneath the surface in the Groningen gas field, the sandstone gradually compresses resulting in surface subsidence. The sandstone layers situated along the fault lines can move more quickly causing minor earthquakes. These fault lines lie on the north-eastern and southern areas of the gas field, where 60,000 people live, which is where the earthquakes are concentrated.[9]

The first officially registered earthquake was in Assen in December 1986, although there were quakes in 1976, 1981, and 1984. As early as 1963, engineer Willem Meiborg (nicknamed 'Willem Beton') warned of the dangers of soil subsidence due to gas production; and Meiborg inspired the activist group Willem Beton. Meent van der Sluis (1944-2000), social geographer and then member of the Drenthe provincial council, was the whistle blower who made the link between earthquakes and gas production. Van der Sluis kept records, visited sites and farmers, and documented seismic activity on a large map covered with increasing numbers of drawing pins. In *The Dutch Drowning Syndrome* (1992), he warned of impending damage to the Wadden Sea area when permission was granted to drill there. NAM rejected any suggestion of a link between earthquakes and gas production and dismissed Van der Sluis as a geographer 'not a geologist'. NAM claimed to be fully au fait with fault lines and gas production; and the KNMI, centre for national seismology, corroborated their stance.[10] Between 1986 and 2013, more than 1,000 minor earthquakes were recorded which, together with several tremors of greater magnitude, led to a gradual – if grudging – rescinding of the earlier official position. Media attention for the damage and protests against the NAM's eventual compensation measures, ultimately led to the government decision in 2019 to 'turn off the gas-tap' – faster than initially anticipated.

As public debate about earthquake damage grew, the Meent van der Sluis Foundation was established in 2015 with the sole purpose of showing and telling the other side of the gas story celebrated by the Slochter Molecule. The aim, in so doing, was to rehabilitate the whistle blower's reputation. The Foundation, together with a group of artists, had an idea for a monument. They approached Groningen artist Karel Buskes who, in consultation with the commissioners, came up with the figurative image of a brick standing in a field. The work was informed by theatre design: the simple, powerful, abstracted form is illuminated at night from within transmuting it into a light sculpture. The cracked brick is easy to read by those who have experienced damage to their houses. Public awareness of the Groningen drama has certainly increased through such publications as *De Gaskolonie. Van nationale bodemschat tot Groningse tragedie* and *Gronings goud*; and the successful play *GAS*, by toneelgroep Jan Vos. The pedigree of artistic response can be traced back to the *Waddenkunstkring*, established in 1997, which tried to capture public attention for the Wadden area through art.

This brief account of how art was mobilized by a foundation committed to opening up and questioning entrenched positions sets the scene for an imagined conversation between the protagonists.

Monumental conversation

Imagine driving across the east-west axis of the Groningen gas field on the A7 and meeting the monumental molecule and brick along the way. What might they say to one another – and to us?

Slochter Molecule: 'Hi! Look at *Me*: I am **Methane**! I've got five pearly balls: four soft silver hydrogen atoms and, at my heart, is the midnight blue carbon atom! No, I'm not a Jeff Koons look-alike; I'm a home-grown Marc Ruygrok work and I've been standing here on the median strip for more than ten years. I was given as a present by the NAM to the people of Groningen, in 2009, to commemorate a half century of lucrative extraction. I was unveiled by Her Majesty Queen Beatrix of the Netherlands and the top Shell men. So never confuse my Methane form with Jeff Koons' frivolity – even if the odd CEO driving at 130 km per hour in a Jaguar thinks he's seen a gambolling bambi somewhere on the road.'
Ethnographer: 'But why *Methane*?'
Slochter Molecule: 'Well, methane is *the* most important element of natural gas. I make visible the tiniest molecular structure of the key domestic energy source and basis of Dutch prosperity since 1959. I am not just visible in some static, boring museal way: I am out there, public art, visible only at

speed – looming over the horizon, flashing past in the blink of an eye, and disappearing behind you again as you accelerate over the hill after the exit to Slochteren. Conceptually speaking I am in motion, too. My movement out of the ground and through the vast infrastructure of pipes echoes the movement of traffic along the road. I go to the essence of the Groningen gasfield as you speed over it! **I ♥ Methane:** don't you ever forget it! Slochteren is where it all started with the Shell and Exxon Mobil's exploratory gas well in 1959, which later became one of the largest gas fields in the world. Carry on with your journey and busy lives, but remember me as you light the gas to cook, stand under a hot shower, or turn up the central heating on a cold winter's morning'.[11]

Atomium, Brussels: 'Hello: this is Atomium joining the call from Brussels. Obviously, your form, Slochter Molecule, is substantially indebted to *me*. It was I – for Iron – who inspired André Waterkeyn to create the nine spheres of an iron crystal, 165 billion times larger than life as the flagship for Brussels 1958 World's Fair. I embody technical/ scientific progress – put to peaceful use in atomic energy. I'm a "monumental structure halfway between sculpture and architecture where the cube flirts with the sphere, a remnant of the past with resolutely futuristic looks, a museum and exhibition centre: ...[I am] both an object, a place, a space, a utopia and a unique emblem in the world which – ultimately – escapes all form of classification ... [I was] restored in 2006, and some refer to me as the most Belgian of monuments"'.[12]

Slochter Molecule: 'You know, if there is one thing that distinguishes the Dutch from the Belgians it is speed. That is, indeed, the difference between Iron and Methane: you speak like a lost surrealist trying to persuade the population of hard reality in a land of multiple truths. I'm not denying that reality looked different in the aftermath of the Second World War and that reconstructing war-torn Europe led to an admirable upsurge of prosperity, optimism, and confidence in the future. But really: 50 years later it's clear that Methane has the edge on Iron, and nuclear power. Just look at the way the Iron Lady got the sack in Great Britain: that was a semiotic diagnostic if ever there was one. No, Atomium, your marvellous form may have inspired my art-worker, but your ideology and atomic energy are things of the past. My location, which is unapproachable on foot and preferably viewed at 130 kmph, says it all when it comes to the way your northern neighbours look at progress. We don't shout about it in the middle of The Hague or even Amsterdam. We go to the local level of truth and assemble a shimmering spectacle of 50 years Natural Gas Extraction that has Enriched our lives in the most literal sense of the term. Get Real, Atomium!'

At that moment, the ground shuddered beneath Methane and across the southern Groningen gas field: the beautiful bricks of twenty monumental farmhouses, dwellings, and barns were rent apart and their walls revealed fissures, cracks, like the dark crevices of mouths once graced with teeth. From the wings of the motorway, a monumental broken brick stumbled across the ditch and into the adjoining field.

Other Monument: 'Ha [said the brick, once in place]. Now it's time to hear the other side of the story. We Bricks connect Groningers to their ancestral past. There were brick ovens in Loppersum as early as 1537, and the remains of medieval brick ovens go back to the 13th century. And, of course, the Romans brought bricks here originally around 300 AD. My dimensions are those of the typical Groningen brick (1: 2: 4). We Bricks moved with the times, though: our production got mechanized from 1850 onwards. No wonder that foreign visitors wax lyrical about the qualities of Groningen brick and the craftsmanship of Dutch vernacular building more generally.[13]

But here I stand: you could see me as the molecule of the house. I'm somewhere in between you, Methane (blown-up from the tiniest scale), and the NAM (which is the biggest player round here). I'm on a human scale; everyone gets what I stand for. I'm not aiming for smooth perfection, even though weathering steel is great material and gives me a beautiful skin. It's my laceration that's my message: you can see it at night when, illuminated from inside, I transform from brick into a light object. You can also find Meent van der Sluis' profile in the jagged edges of my gash: so although I'm an earthquake monument in the shape of a brick, I also bear a portrait of the man who dared to warn of the dangers of gas extraction, at a time when that was not done.'

So there they stand, together, though separated by the changing insight of a decade: the unapproachable, frolicking Methane, witness to the promise of endless prosperity; and the Molecule's own Other – the sentinel Brick who, upended and fissured, reminds us of what is going on under our feet and illuminates the shaken structures of our houses, monuments, and other buildings.

Other Monument: 'Just one last thing: I'm not a protest monument. My artist isn't a manifesto-writer or an activist. I'm actually more interested in *talking* to you, Methane, than I am in iconoclasm. And more excited by the impact on the same driver who, leaving Bad Nieuweschans for Groningen, first passes you and 6.7 km later me! Or *vice versa*.'

Slochter Molecule: 'Well, lucky you: you didn't get daubed yet – as I have been, several times, and had to be cleaned off. *So* demeaning! I'm a work of art – and you want to turn me into a monument! But just you wait: Corten steel is a great surface for graffiti artists.'

Other Monument: 'Now, there's the thing, Methane. It could always happen, of course. But unlike you, who was designed by an artist from Den Helder, fabricated in Berlin, and unveiled by HKH/ Shell, *I'm* the work of a Groningen artist. I commemorate an Assen visionary. And I was made by students from the Metal and Shipbuilding Training (MSO) in Groningen – many of whom had never been involved with art before. Even if some saw the NAM contribution as a pay-off, support from the Mondriaan Stichting clearly vindicates my art value. I was unveiled by the former Finance Minister who, nota bene, became chairman of the Dutch Safety Board a couple of weeks earlier. The family of Meent van der Sluis were honoured guests at my inauguration. And the NAM have admitted they were wrong. These developments alter our understanding of the past.'

Slochter Molecule: 'So we're in this together? Not in hindsight, not in the rear-view mirror, but here and now in a world where the maximum speed limit is back at 100, the gas tap is being turned off, the gas burner has been ousted from the Dutch historical Canon, and we are looking for alternatives!'

Other Monument: 'Yes, and there's certainly room for more art along this road.'

Atomium: 'Guys: word is that – by fission or by fusion – atomic energy may be making a comeback, in combination with wind and sun energy.'

Slochter Molecule: 'Hmm, were you thinking of a replica of yourself in Groningen? Take it from me, the vast infrastructure created for *my* transportation is being repurposed for hydrogen – which can be used as a vehicle for energy generated by windmills. That gives pause for thought, doesn't it?'

Other Monument: 'Wait, everyone! The Slochter Molecule and I already form a pair: and you, Atomium, have clarified the link between yourself and Slochter Molecule. We're all time-based artworks of sorts, conversation pieces that, in occupying space, interrupt the flow of time, make people think about the complexity of our entangled lives, and slowly change minds'.[14]

Conclusion

Imagining a conversation between two monuments is a way of 'thinking through things':[15] attending to things, as they emerge in specific settings, according to how they are presented. In this case, the 'other monument' was designed as an explicit 'answer' to an older artwork which it (the new one) transformed into a monument by self-designation and implication. Being transformed into a monument in this way demonstrates the agency of art in redefining the field of memory. The transformational potential of this 'answer' prompted me to explore positions and possible changes of position through a monumental conversation. Monuments, as artworks, speak through their scale, form, colour, dimensions, materials, manufacture, location, and aesthetics, as well as through juxtaposition and proximity to other artworks, about memory and shifting views of the past. Unlike the historical statues that were toppled in the wake of political protests in 2020, the two Groningen monuments embody natural gas and a brick, respectively. Yet given the historical and political circumstances attending its installation, it is easy to imagine the Slochter Molecule as a narcissistic and uppity character: ebullient form, blown-up scale, lightweight aesthetic, external illumination by night, inaccessibility, and top-down history tend that way. The Other Monument, by contrast, takes a familiar form, scaled up to expose a fracture, and illuminated from within at night to reveal a portrait of the man who challenged the narratives of vested interests that denied any connection between earthquakes and gas extraction. We warm to the Brick for exemplifying how putting up new statues rather than tearing down old ones can provide a handle on the complexities of history.[16] The monumental landscape in Groningen shows how opening the possibility for dialogue among protagonists occupying different historical, political, and emotional positions, matters.

With grateful thanks to Trijni van der Sluis, Piter Bergstra, and Karel Buskes

Suggestions for further reading

M. Brandsma, H. Ekker, R. Start. *De Gaskolonie: van nationale bodemschat tot Groningse tragedie.* Groningen: Passage, 2016.

D. Lowenthal. *The Past is a Foreign Country.* Cambridge: Cambridge University Press, 1985.

J.E. Young. 'The Counter-Monument: Memory Against Itself in Germany Today'. *Critical Inquiry,* 18(2) (1992), 267-96.

The challenge of living on renewable energy

Anton E.M. van de Ven

Introduction

To a large extent, the quality of our lives is based on the ready availability of physical power. We may not give it much thought, but we require such power for nearly all the things we do, from building our houses and highways to growing our food and travelling places. At present, almost all of our activities run on fossil fuels but coal, oil, and natural gas are non-renewable. Furthermore, burning such fuels releases CO_2, which causes dangerous climate change. To counteract this, it has been proposed to switch to renewable forms of energy, such as solar or wind power, and thus hopefully meet the goals of the 2015 Paris Agreement. This agreement asks us to keep the increase in global temperature to below 2°C above the pre-industrial level. The transition to renewable forms of energy, now under way, is a major challenge which should not be underestimated. It is important to know which kinds of renewable energy are realistic options. One frequently reads that there are 'huge' amounts of solar and wind energy to be had but, as for instance Sir David MacKay has emphasized,[1] here we need relevant numbers, not meaningless adjectives. Educated citizens should not only be literate but also numerate, so they can judge such claims for themselves and make informed choices. More generally, numbers are needed in debates but are often missing.[2]

In this essay, we shall start from a historical perspective to show that living on renewable sources of energy is not a recent invention. Not so long ago, our activities relied on what little power we could gain from wood, wind, water, sun, draught animals, and our muscles. No sooner were more powerful sources found or we abandoned the old ways. Still, past energy transitions easily took a century and always worked towards better sources of power. This is not true of the present (re)turn to renewables: modern wind turbines are more efficient than old mills, but the wind still offers little power and it remains intermittent. Much the same is true of any other renewable source of energy. Hence, for any mix of renewables, country-sized facilities will be needed to provide for the present levels of energy use, all the more so because in this century, the world population is predicted to reach ten billion and all these people hope to lead better lives. Our world will be utterly transformed by the ongoing energy transition, certainly if it were to be a return to renewables only. We will argue here that it would be very challenging to live on renewables only. Other carbon-free sources of energy, most likely nuclear energy, will be needed too. At the same time, we must increase the energy efficiency of all processes, invest in new ways of storing renewables, and reduce per capita power usage. Finally, all these efforts may well fail if we do not also change our hearts and minds about what it means to lead fulfilling lives. Clearly then, tackling the issue of the energy transition needs to involve the humanities and social sciences.

'Renewable' versus 'sustainable'

Energy is the ability to do work and is measured in joules, J. Power is the *rate* at which work is done, expressed in watts, W, with $1\ W = 1\ J/s$. Lifting a book by one meter requires you to do roughly one joule of work. If you do this in one second, your power was one watt. Adults can briefly sustain a power of a few hundred watts. In Europe, the average total power used per capita is roughly 5 kW (5000 watts). In common parlance, and also in much of the literature, renewable energy and sustainable energy are used as synonyms, but here we want to make a distinction. A source of energy is *renewable* if it is replenished within a human lifetime. We may for instance use the warmth of the sun today to dry the laundry and yet it will shine again tomorrow. Fossil fuels, stored solar energy from the distant past, are not renewable. Oil production, depending on whom one asks, is expected to peak some time in this century. These examples show that renewability is a simple and well-defined notion.

Far more complex and ill-defined is the idea of *sustainable* energy (in fact, sustainable *anything* is complicated and not so clear). We will call a source of energy *sustainable* if it is renewable and in addition is used efficiently, avoids bad environmental effects, respects social justice, and is economically feasible (some add the requirement that it not violate their aesthetical preferences). Thus, some kinds of renewable energy may not be sustainable. Controversial examples are easily come by. In the past, humans relied mostly on burning biomass – wood and other materials – to provide heat and energy. Trees and other plants grow within a human lifetime or less, so this is a renewable form of energy. In order to accelerate its transition to renewables, the Netherlands is now substituting some of the coal burned in its power plants by wood pellets. However, on several points the use of wood pellets may well be deemed non-sustainable.

It is important to evaluate a proposed renewable power source according to the criteria above to see if it qualifies as a sustainable source. However, doing so runs the risk of disqualifying some options out of hand – wind turbines, it is said, are ugly – and simply continuing our present reliance on fossil fuels. We should rather start by asking if we can run our society on renewables only and first study physical feasibility. After all, we can always print more money, change our ethics or our sense of aesthetics, but the laws of nature are immutable. Although we might yet change our minds about some of these laws, it would not be wise to count on this to solve our energy problem. In particular, the laws of thermodynamics firmly hold and *perpetuum mobiles* simply do not exist. Our query requires numbers on present energy use and proper estimates for the possible yield of each renewable energy source. Should we discover that renewables add up, we may next impose the further demands of sustainability: ethics, aesthetics, environment, and economics. However, before judging renewables based on their physical feasibility, let us see what we can learn from past energy transitions.

The transition to renewable energy from a historical perspective

Humankind seems to learn few lessons from history, but this is a case where we might do so. Indeed, energy transitions of the past have been studied in detail.[3] Wind turbines and solar power towers are modern inventions, but the exploitation of these energy resources is not new. Once upon a time, not so long ago and in a country not so far away, people lived on renewables only! They kept horses and built windmills and waterwheels to power their world. Although entirely utilitarian, many a mill became the subject of a painter's canvas in the works of artists such as Ruysdael, Gabriël, and Mauve. Like them, we may have a rather romantic view of such old mills, but they were designed as machines to deliver power. No sooner was a new source of energy discovered or we switched and generally without regret. These transitions did not require government programmes or subsidies. The market did its thing because the new sources were 'hotter' (more energy dense). Even so, the transition from wood to coal took at least a century.

As MacKay has put it, the Industrial Revolution in the UK was a coal revolution and the amount of coal under Britain was about equal to the amount of oil now under Saudi Arabia. UK coal production peaked in 1913, then steadily declined, and ended after the 1994 privatization of the mines. This was foreseen in 1865 by economist William Stanley Jevons,[4] who wrote in *The Coal Question*: 'I must point out the painful fact that such a rate of growth will before long render our consumption of coal comparable with the total supply. In the increasing depth and difficulty of coal mining we shall meet that vague, but inevitable boundary that will stop our progress.' For the Netherlands, it is interesting to note that it was not our windmills but peat that constituted the main source of energy in the seventeenth century. The roughly 3,000 windmills on average yielded about 10 MW (megawatts), but the exploitation of 1,700 km² of peat bogs sustained a hundredfold greater power.[5] After the peat ran out, the Netherlands relied briefly on coal, imported oil and, post 1960, relied on the natural gas found in Groningen (see the article by Mary Bouquet in this volume). The precipitous end of Dutch gas production is now being brought on by the impact of minor, but damaging, earthquakes in the region. If not for these quakes, the Dutch would surely have continued the use of this energy source. The Netherlands, and with it most of the EU, is now turning to Russia, Norway, and the US for its gas.

Other countries have their own energy stories to tell. For instance, in the eighteenth century, almost every stream in France had a watermill which, all together, powered the industrial revolution. Having little coal, France turned with success to nuclear power as a workhorse after the Second World War. Belgium and Germany long relied on coal and later in part on nuclear power. After the 2011 Fukushima incident, the Germans opted for their *Energiewende*, investing massively in wind and solar energy and shutting down all its nuclear plants, even though the lesson of this incident should have been that such plants should not be built in earthquake zones. This is bad news for Japan but good news for most of the EU. However, there is ever stronger resistance to more wind turbines to the point where further growth in this direction is stalling. Belatedly, Germany has begun to turn off its coal-fired power plants too. Combined with new laws against more wind turbines this may well lead to a German domestic energy shortage.

Looking at these past energy transitions, we note that they have in common that they took up to a hundred years. Also, instead of giving up on any particular source, it is the case that all countries now use a mix of all energy sources and, notwithstanding the recent rapid growth of wind and solar energy, fossil fuels continue to provide 85% of the world's energy.[6] We should rapidly reduce this reliance on fossil fuels. One option may then be a return to renewables.

Renewables only?

The sun shines for free but has low power density. In that sense the present energy transition is *not* towards a better source. On average, we get 1-20 watt per m^2 of land or sea. To provide then a billion watts of electricity – the needs of a city – requires circa 50 km^2 of solar cells at 20 per cent efficiency. Furthermore, most renewables are intermittent: wind turbines run about 30 per cent of the time and the sun does not shine at night. If we insist on energy on demand then we need to store it. This adds considerably to the cost and environmental impact of renewables. Also, just like fossil fuels, renewables are patchy and not equally distributed over the planet: there is a lot of sun in the Sahara, but not many people live there. Besides benefits one also can foresee geo-political issues, not unlike those now caused by our dependence on oil, should we start producing solar power for the EU in North-Africa.

Even with modern technology and the addition of geothermal and tidal energy, living on only renewable sources will be difficult. The problem is not just that such sources are more diffuse, intermittent, and costly. It is also that the global human population has increased sevenfold since the industrial revolution and will increase even more. Furthermore, at least in the West, lifestyles now demand much more energy (per capita use has increased tenfold since 1800).[7]

An interesting study on the feasibility of living on renewables in the UK was undertaken by Sir David MacKay. He made numerical, rather optimistic, estimates for the possible contribution of each type of renewable source. The UK is blessed with wind, tidal, and wave energy sources. Adding it all up, MacKay found it to be just conceivable that the UK could live on its own renewables. However, this would require a complete overhaul of the British landscape and way of living. Adding typical societal and economic factors, MacKay concluded the UK cannot do so.[8] One concern, not just in the UK, is the perception that wind turbines and solar installations are ugly. Here, educating citizens, making them co-owners of such power plants, and the involvement of the arts are important. In Portugal and Australia, for instance, artists collaborated with their local wind industry to turn wind turbines into works of art.[9]

Although the details differ, the same conclusion MacKay reached for the UK can more or less be drawn for other locations. There are exceptions, as for instance shown by the island of Samsø in Denmark.[10] For its electricity and most of its heating, Samsø is now largely self-sufficient. The island aims to be fossil-free, also for transportation, by 2030. Such inspiring examples demonstrate that, at least in places with much wind power and small populations, a new way of living is feasible. It is doubtful though whether this would work in more densely populated settings. Another option we should look at then is nuclear power.

Nuclear power?

The nuclear accidents of Chernobyl and Fukushima are well-known and scary. Even as most nuclear power plants operate safely, there have been delays and associated cost overruns in construction. Add to this the issues of nuclear waste, decommissioning costs, and nuclear arms proliferation and it seems clear why many would rather not grant nuclear power a role. Still, nuclear fission, based on the splitting of heavy atomic nuclei, is one of the few ways now available to us capable of generating much power with very low emissions of CO_2, which is the crucial issue. By the middle of the century, it is expected that commercial nuclear fusion, based on the merging of light atomic nuclei, will become available. Indeed, the International Thermonuclear Experimental Reactor (ITER), now under construction in Cadarache, France, offers hope that this form of sustainable nuclear energy will be available after 2050. Per kilogram of fuel, fission reactors offer up to a million times more energy than fossil fuel power plants and even better yields are expected for fusion reactors. A 1 GW (gigawatt) nuclear power plant takes just a few hectares instead of the large swaths of land or sea required for the same yield by means of renewables. Thus, nuclear power, whether based on fission or fusion, combined with new forms of agriculture would allow the rewilding of large spaces as advocated by the ecomodernists.[11] In MacKay's words, the nuclear waste issue is a beautifully small problem, certainly when we compare it to the 43 gigatons of CO_2 which humanity emits each year. More importantly, new generations of nuclear fission reactors have been developed which are much safer and produce less, shorter-lived, nuclear wastes. Present- day reactors rely on uranium, but we can use the more plentiful element thorium as a fuel too. The thorium fuel cycle produces less waste and cannot readily be used to make nuclear weapons. The trusted UN's Intergovernmental Panel on Climate Change sees an important role for nuclear power in all its future energy scenarios if we want to meet the targets of the Paris Agreement.[12] Also, some famous environmental activists have changed their mind about the topic. As James Lovelock put it: 'I am a Green, and I entreat my friends in the movement to drop their wrongheaded objection to nuclear energy'.[13]

Democratic societies may choose to reject nuclear power, but only after having taken account of all relevant facts and risks. One should realize the grave danger caused by a continued reliance on fossil fuels and the limited time we have to act. Of course, citizens in high-income nations should feel free to instead change their lifestyles and use much less energy.

Conclusion

Our discussion has reminded us of the extended time and real effort required for past energy transitions. One should note the scale of the present energy transition to renewables. Given that we are moving to 'less hot' energy sources, it is an uphill battle which will require concerted and sustained government and citizens' support. Still, in the EU we are already seeing that wind energy can grow without subsidies. If we also remove government support for fossil sources and continue to educate and involve our citizens, the energy transition may run by itself. Nevertheless, given our present way of living, it is not possible to rely on only renewable energy. If we continue our ways, renewable facilities will have to be the size of a country. Given the urgency to get off fossil fuels and to allow a decent living standard for ten billion people, nuclear fission would be an interim solution and nuclear fusion one for after 2050.

Sustainability connects humans to nature, joins us to our ancestors and our descendants, and relates global to local issues. In his *Ethics* (Part II, Proposition XLIV, corollary II), Spinoza wrote that it is in the nature of reason to perceive things from the perspective of the eternal (*sub specie aeternitatis*).[14] I believe that living sustainably requires us to take a wider perspective in both space and time. It asks us to care not just about living in the here and now but also about those who will come after us on this planet. Learning to live in a sustainable way on our Earth is possibly the greatest challenge facing humankind. Urgent issues, from climate change to loss of biodiversity and population growth, are interconnected and cannot be dealt with one at a time. We also can no longer move to new continents, but science, supported by the arts and humanities, can still help to solve our predicament. Unlike past transitions, this can only work if we are also willing to rethink and truly change our way of living.

Suggestions for further reading

D.J.C. MacKay. *Sustainable Energy – without the hot air.* www.withouthotair.com, 2008.

V. Smil. *Grand Transitions: How the Modern World was Made.* Oxford: Oxford University Press, 2021.

V. Smil. *Numbers Don't Lie: 71 Stories to Help us Understand the Modern World.* London: Penguin Books, 2021.

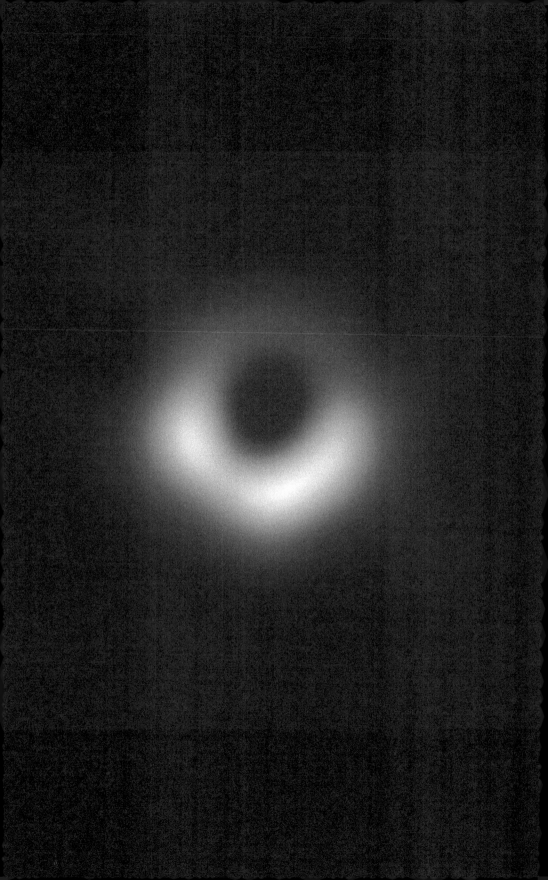

The earth as an observatory: Team work in science

Filipe Freire

Introduction

On 10 April 2019, a consortium of scientists released a very special and long awaited photo that made front page news across the world and lit up Twitter feeds.[1] The photo was the very first image of a black hole, providing proof of the existence of this astronomical object first named by the American physicist John Wheeler in 1967. The resolution required to produce the picture was equivalent to the sharpness necessary to spot an orange lying on the surface of the moon. It was made possible by scientists in 64 institutions around the world working together with data collected from radio-telescopes in the USA, Chile, Mexico, Spain, Greenland, and Antarctica. The project, which is still continuing, is known as the Event Horizon Telescope or EHT (https://eventhorizontelescope.org). What has already been accomplished in this project is not only a great scientific and technological achievement, but also an example of international cooperation, involving a large group of scientists unified by a common method, passion, and goal. In many ways this landmark result illustrates twenty-first-century scientific cooperation at its best.

So what is it about the notion of black holes that has captured the public imagination and what was actually involved in producing the headline-making image? In the first part of this essay I will introduce you to some of the science behind black holes and the collaborative processes and technology involved in the EHT. In the second part of this essay I give a brief description of an outreach project conducted at University College Utrecht, which introduces students to real teamwork in physics and lets them analyse cosmic ray data, collected by detectors on several locations in the Netherlands and abroad. Cosmic rays are tiny particles that come from outer space and constantly shower the Earth.

The black hole: the great unknown?

It could be argued that the black hole has become the metaphor for 'the great unknown'. In popular culture it is often associated with its trapping power or with being sucked into funnels as in a whirlpool. Black holes have also featured in movies, most recently in Christopher Nolan's well-received 2014 film, *Interstellar*, where the black hole was given a name, *Gargantua*, after the famous hungry giant from Rabelais' sixteenth-century novel. In a book about the film's science the physicist Kip Thorne, who was one of its main scientific advisors, suggests that much of *Interstellar*'s science is at or just beyond today's frontiers of human understanding.[2]

Matter and gravity are useful starting points to explain the basic feature of a black hole. Matter attracts matter because gravity is an attractive force. So when we have an extremely large amount of matter in a small region, nothing can stop this matter from 'falling' into its centre. The density of the matter keeps growing, it completely collapses into itself and eventually forms a black hole. This happens when it creates a region around it from where nothing can escape its gravitational attraction and falls into it. If we were to compress all the mass of the Earth into a sphere with a radius of 1 cm it would become a black hole.

The idea that it was possible to compress so much matter into such a tiny sphere was so absurd that Einstein and many others thought that this could not really happen. After all, we have stars and planets, which provides evidence that there should always be something to stop matter from completely collapsing into itself and forming a black hole. Surely *nature* has found a way to prevent this from happening. Between the publication of Einstein's theory of general relativity in 1915 until the 1960s, this was what most physicists and astrophysicists thought. The game changer came in 1967 with the discovery of pulsars, fast spinning neutron stars. These objects or stellar bodies have about 1.5 times the mass of the sun within a radius of just 10 km. Not black holes as such, but close enough being one that the idea they could actually exist had to be taken seriously. A neutron star is already far smaller than what was thought possible. This discovery marked the start of the fascination with black holes: something that should not exist but actually does. For theorists the black hole phenomenon became a test bed to search for a more fundamental understanding of the laws of physics and the structure of space-time and matter.

In 2021, we know that there are many black holes in our universe. In fact, there is very strong evidence that there is a very massive black hole at the centre of any large galaxy, including our own Milky Way. Currently, around 100 billion (100,000,000,000) galaxies have been mapped. These black holes have masses ranging from millions to billion times the mass of the sun. The black hole captured by the EHT is one of these super-massive black holes, which is at the centre of the M87 galaxy and represents the strongest visual evidence of their existence to date. It has almost as much mass as the entire Milky Way.

Besides their central black hole, each galaxy has many of the so-called stellar black holes, with masses ranging from five to fifteen times the mass of the sun. These are the core remnants of a star estimated to have had a mass of about ten to twenty times that of the sun. Less than 1 per cent of all stars are expected to end up as a black hole. In our Milky Way this amounts to about ten to 100 million stellar black holes. Clearly, there should be many stellar black holes, but it is difficult to observe them. Unlike the super-massive black holes they do not attract vast amounts of matter that light up their surroundings (see caption to the image).

How can we detect stellar black holes or estimate their numbers from observation? The cosmic collision between a black hole and another dense object, like another black hole or a neutron star, produce waves on the structure of space-time that can be detected by gravitational wave detectors. The first detection of gravitational waves was in 2015 by the Laser Interferometer Gravitational-Wave Observatory (LIGO) (www.ligo.org). Since then LIGO, together with the Europe-based VIRGO detector, (www.virgo-gw.eu) have detected dozens of black holes (Cho, 2020).[3]

The making of the black hole image

The image of the super-massive black hole first shared with the world in 2019 was a true tour de force that could not have been achieved without modern technology and large-scale team work. It involved a wide range of experts working together: engineers, astronomers, physicists, and computer scientists with combined technical expertise in instrumentation, data collection and processing, image reconstruction algorithms, modelling, and simulations. It also involved good communicators – who were able to explain both the achievement and the significance of the image.

What at first looks like a blurry snapshot of a campfire to many of us, is in fact a unique image, the production of which required overcoming many technical challenges. The main challenge was to achieve an extremely high resolution – as noted earlier, sharp enough to spot an orange on the surface of the moon. The solution was to use very long base interferometry (VLBI), a technique that combines the radio-waves captured by an array of radio-telescope antennas. The array behaves as a single earth-size radio-telescope when the longest possible distance between the antennas is of the same order as the diameter of the earth. The resolution is also increased by observing at the shortest possible radio wave-length. The EHT astronomers choose to observe on wave-lengths of 1.3 mm or shorter. This technique had not been used before for such long distances and short wave-lengths, and several years of preparatory work had to be done.

But the challenges did not end there. The data from all the radio-telescopes in far-apart locations needed to be transmitted and centrally stored. Only then did the actual work of producing an image start, and it took two years. Data collected at different stations and at different times needed to be synchronised to ensure that the information on each image corresponded to the black hole at a given time. The image had the required resolution thanks to the earth-wide array of radio-telescopes, but the data collected was not sufficient to reproduce the entire image directly, because the radio-waves are only collected at the telescope locations. For this stage, the EHT collaboration had to develop image reproduction algorithms to complete the final image, and had four teams working independently on this. Clearly a massive team effort.

One of the fascinating aspects of the project, which is ongoing, is that it uses the earth as an observatory, with antennas in different locations around the world, and it involved hundreds of scientists and technicians working on data collected from the black hole at the centre of the galaxy M87. This work also represents something larger but somewhat similar to what was experienced when humans 'conquered' space and travelled to the moon. Back then, in the late 1960s, we were humbled by the image of our blue planet as seen from space. Now, we are humbled by an image, showing a vast zone of our universe, several times larger than our entire solar system, which is completely inhospitable. A stark contrast to the amazing life-giving planet we inhabit.

We may think that what used to be *symbols* or *myths* about black holes, can now be 'captured' in an image. But in this case, the image by no means demystifies what a black hole is. For instance, we still know very little of what goes on inside a black hole, and we do not yet fully understand how the radio waves in the accretion disk are produced. Much more science will certainly be generated by these mass cooperative endeavours.

Working on big data in the neighbourhood: the HiSPARC experiment

Whether we are trying to better understand the physics of a black hole or detecting gravitational waves, we need to analyse large amounts of data.

Collecting data on such a large scale does not just enhance cooperation between scientists across the world, it also has great potential for acquainting a broad public of lay scientists, including secondary school students, with experimental/observational physics. The HiSPARC experiment, which stands for High School Project on Astrophysics Research with Cosmics, is an example of a project that uses multiple detection stations, and links research institutes, universities, and secondary schools. This project has been designed as a network to detect and monitor the behaviour of cosmic rays, and is coordinated by the Dutch National Institute for Sub-Atomic Physics (NIKHEF). Cosmic rays are high-energy particles that reach the Earth from very faraway places in the universe.

Currently there are cosmic ray detector stations in several cities in the Netherlands (Nijmegen, Amsterdam, Eindhoven, Enschede, Leiden, Utrecht, and Groningen) as well as in Denmark and England. Three detector stations are located at University College Utrecht, where there is also a weather station. As with the EHT collaboration, data collected by HiSPARC first needs to be stored in a central location and all detected events are timestamped and localised using GPS antennas linked to each station. NIKHEF hosts the computers where the data is stored and made accessible for analysis.

The HiSPARC detectors collect data twenty-four hours per day. At each participating school or university, the physics teacher and students are responsible for maintaining their stations. The detectors are kept inside ski-boxes on the flat roof of a building and their activity is monitored by an electronic box and a pc. Their maintenance does not require a high level of technical knowledge. In turn, students can work on small projects with HiSPARC data and/or associated technology. For example, at University College Utrecht, students have been developing programming code to study the monthly and annual cosmic ray flux modulation.

What is special about the cosmic rays being researched in HiSPARC is their energy, which is equivalent to the energy transmitted when serving a tennis ball but concentrated in something as tiny as a proton. Why do we need detector stations in multiple locations? *Cosmic ray showers* are produced when a high energy cosmic ray enters the atmosphere of the earth. They are made up of millions of particles. The showers produced by high energy cosmic rays have enough energy to reach the sea level and these are the ones we detect with the HiSPARC detector stations. They are spread over a couple of kilometres and are detected by several stations. The data collected from one of these showers is then analysed in order to determine the energy and the sky location the cosmic ray originated from.

Data from at least three stations is required in order to search for the origin of the cosmic ray. To have detectors in multiple locations also increases the probability of observing the most energetic cosmic rays.

You might wonder why we care about these cosmic ray events. Well, with our increasing dependency on new technology like smartphones, it is important to know more about these showers since their high energy could cause electronic devices to malfunction. Fortunately, when cosmic rays reach the earth's surface, they are not harmful to humans, but the primordial high energy cosmic ray, as it is found in space, can be and is a known risk for cosmonauts in the International Space Station (www.nasa. gov/mission_pages/station).

A special and rare case is when the amount of energy in some of the primary particles in the cosmic rays is so high that we do not know of any process that can explain how they were produced. We call these *ultra-high-energy* cosmic rays. In an area of one square kilometre, we would only expect to detect one of these per century. In order to detect one per year you would then need to distribute detectors over an area of 100 square kilometres. The largest experiment to detect cosmic rays to date covers an area of 3,000 square kilometres in southern Argentina. At present we still do not know where the ultra-high-energy cosmic rays come from and more data needs to be collected. One of the most likely possibilities is that they are produced by super-massive black holes as the one in the image produced by EHT. HiSPARC could detect these rare events because of the large area covered by its stations.

Conclusion

In this essay I have reflected on the collaborative nature of a scientific endeavour using as an example the years of work and huge range of expertise and skills which led to the first ever image of a black hole. This was a slow and complex process at a time when we expect answers and results by a single click on our smartphone. Collaboration is also at heart of the scientific method. For instance, one of the main points of dispute in projects, where large amounts of data is analysed, such as EHT, concerns methodology and interpretation of data. Indeed, disagreements are actually welcomed as they force each individual and team involved to be even more systematic and rethink each step and decision. This systematic thinking is also part of HiSPARC, one of the main goals of which is to introduce students to the team work and rigour that is required in experimental or observational projects in physics and astronomy, in addition to the technology itself.

In conclusion, there are good reasons to believe that we are now entering a golden era of black hole observation. For current and future researchers the development of the technology to produce black hole images, as well as the detection of gravitational waves, offers many possibilities to find the answer to questions such as: How many black holes are there actually? How do super-massive black holes grow so big? This is indeed a great time to consider joining this exciting scientific community.

Suggestions for further reading

M. Bartusiak. *Black Hole: How an Idea Abandoned by Newtonians, Hated by Einstein, and Gambled on by Hawking Became Loved.* New Haven, CT: Yale University Press, 2015.

K. Thorne. *The Science of Interstellar.* New York, NY: W.W. Norton & Company, 2014.

C. Impey. *Einstein's Monster: The Life and Times of Black Holes.* New York, NY: W.W. Norton & Company, 2018.

The First Assignment

Markha Valenta

For Barry O'Connell

It was a deceptively simple assignment: 'Write about a time that words made a difference'. The simplicity and the personal question made it like no other assignment I'd had before, more confusing. I'd never thought about words, much less words making a difference. Any more than I thought about breathing.

Of course, the minute you stop breathing, you begin to die. In a way, you could say that about words too. Others ceasing to listen and speak to us means social death. Full isolation from the words of others can drive us mad. In that sense, every word makes a difference. More striking yet are the words that intervene to reshape our lives.

The assignment on words came early on in a course called 'The South as a Literary Landscape'. When I first looked through the course catalogue, the title bounced around awkwardly in my mind. How could literature – a figment trapped between the covers of a book – have a relation to land – tangible earth, gritty, and out there, without a clear beginning or end? And why would the literary landscape of someplace called 'the South' be one to which you would devote a whole course? But the professor was renowned on our campus as one who inspired his students. So I signed up.

Before 'The South as a Literary Landscape', there was only one American landscape that had shaped my life: New Mexico. While I had lived in many places, this was a place whose spaces, textures, and life anchored my family, first around the time of my birth and then again a decade later. When we were not living there, its memory remained a lodestar, a node that held us together.

My mother had come to New Mexico from New York, just out of college with $40 in her pocket. An artist, she responded with all her senses to every space through which she passed. The world was a pulsing flow of textures, colours, patterns. Everything else emerged from this. Art came to my mother as naturally as breathing and was there from the moment she woke. In the summer, this meant painting and sculpting in her underwear, splotches of bright colour on her body, bra, panties. Any errand we had to run might mean my mother picking wood and cloth scraps out of neighbours' curb-side waste to create sublimity. This was deeply embarrassing. But what our pickings conveyed, in a fashion all the more powerful for being unstated, was that art was neither holy genius nor in need of museums; it was simply there, ready to be seen, felt, and made.

To respond in this fashion required a radical openness. I would go into the landscape with that same feeling, lose myself in wanderings between mesas, dust rising in soft puffs across pottery shards and obsidian arrow heads. The land was sensual: the scent of piñon, sage, and crumbling volcanic tuff; rounded adobe walls under piercing blue skies; cottonwoods nestling in the crooks of rivers undulating through the pink-grey dryness. In the midst of that beauty small scraps of history floated by, carried by the palimpsest of people inhabiting the land: Pueblo nations – Navajo, Apache, San Ildefonso, Zuni, Taos, Nambé, Laguna, and others – alongside the descendants of the Spaniards, Mestizos, Hispanos, and now growing numbers of Anglos. A historical and social landscape as unique as the geographic one, older than the rest of the country. A landscape full of blood and dispossession, along with resilience and newness. But one whose story I couldn't tell as a child.

So, when I came into 'The South as a Literary Landscape', I had not yet learned to read landscapes – regional or racial – in terms of their deep histories and politics. My high school in New Jersey had Black and White students in equal numbers, we had read Ralph Ellison and learned of Jim Crow and the racism of the New Deal. Yet none of this prepared me for the intensely transformative fusion of history, politics and aesthetics I would now encounter.

In class, we would read Richard Wright and Eudora Welty, renowned Mississippi writers born within a year of one another in the early twentieth century, into neighbouring but brutally different worlds. Richard Wright's Black childhood meant being bludgeoned into a life shaped to the savagery of both poverty and White terror. Beatings and starvation at home alternated with public lynchings, humiliations, and random bloodletting. Some of the time, Wright lived in Jackson, Welty's birth-city where her firmly delicate, lyric artistry would emerge. Her writing was sensitive to the racial and class injustices around her but was very White in the space Welty had to focus on her craft's aesthetic. Where Wright's Communism and rebellion made his words rough and fierce, her politics were oblique.

Wright never graduated from high school. His success as a writer – the only reason that we were reading him decades later – was because of the voracious intellectual and artistic hunger that drove him to pour out words into harsh, unflinching stories not yet written. Once put to paper, the American literary landscape exploded, then reconfigured itself to their presence.

Alongside these readings, our class engaged the work songs of the enslaved counterposed to the Black songs of freedom and protest from the 1960s. Incantations floating through the air, passed from mouth to mouth, generation to generation. To labour and bind together, to worship and heal, to mobilize as one in the face of evil. Public song brought courage, but also danger. The Black bodies and few White ones that dared to speak, march, and sing across the South would be met by skin-splitting whips, rape, rabid White mobs, dog fangs flashing white and sharp as razors, high pressure hoses, and police truncheons splitting heads.

Here were words that made a difference. The fiction of Wright, the songs of the enslaved, the death-defying chants of the activists. Most potent of all in that course were the living words from the Freedom Summer of 1964 when SNCC ('snick') – the Student Non-Violent Coordinating Committee – recruited Black high school and university students, and northern White ones too, to go onto the highways and byways of Mississippi to the roughest of sharecropper cabins to register Black voters. The year before, Medgar Evers and President Kennedy had been assassinated and four little Black girls blown to bits in the Birmingham Sunday School bombing. Then came 1964, Freedom Summer. This was the year three activists – James Chaney, Andrew Goodman, and Michael Schwerner – would go missing in Mississippi, only to be found by the FBI after 44 days, buried in a dam.

The only reason the FBI was sent in to search for them was that two of them, Goodman and Schwerner, were Northern White Jews. In the course of the six weeks before their remains were found, eight Black bodies of activists were discovered that no police had cared to track down. After the bodies were recovered, the dirt brushed off them, it became clear that the White men who killed James Chaney had given in to a most ferocious drive to torture, denigrate, and erase. Chaney had been chained to a tree, then savagely beaten with more chains. After, or perhaps during, the beating Chaney was castrated. Then finally shot three times. Schwerner and Goodman were made to watch. When Chaney was unchained and fell, Schwerner cradled him. The White men did not like this. Schwerner was called a 'n***r lover'. Then shot through the heart. Goodman fled and was shot on the run.

This history, its stories, were words that poured into me, gripped me, and transformed me. One day I left class on a brilliantly sunny January, the sky an intense New Mexico blue. During the night, it had rained and frozen all at the same time, so that the whole world glistened. Every tree, every branch, every bit of life was covered in a sheer veneer of ice that flashed with brilliance.

The cold air seared my lungs, puffs of steam in and out, and I had no words for any of this. But it was as if my inner eye, my consciousness, my self had cracked open – the endless blue, winter sun and translucent crystalline ice pouring into me and me into them – as I made sense of this world where people fought to vote, to speak, and be heard, where words could get you killed but also free you. A world where being human, being political, and making meaning were brutally dangerous and brutally beautiful, all a part of one life, and that life mine.

That same feeling would come to me listening to Malcolm X some months later, in another class. That class met in the small New England campus church where the English department was housed, all the way upstairs, in a round room under the church tower's eaves. The professor brought in the record with Malcolm's speeches, lowered the needle, Malcolm's voice filling the room. As he spoke, we fell quiet and I lay down on the bench circling the room. This too was a sermon, Malcolm fiercely loving his Black audience, teaching his listeners to love themselves, as Black, as African, as American, as Muslim.

> Who taught you to hate the texture of your hair? Who taught you to hate the colour of your skin to such extent that you bleach, to get like the White man? Who taught you to hate the shape of your nose and the shape of your lip? Who taught you to hate yourself, from the top of your head to the soles of your feet? ... You know. Before you come asking Mr Muhammad does he teach hate, you should ask yourself who taught you to hate being what God made you. [...]
> It's just like when you've got some coffee that's too black, which means it's too strong. What do you do? You integrate it with cream, you make it weak. But if you pour too much cream in it, you won't even know you ever had coffee. It used to be hot, it becomes cool. It used to be strong, it becomes weak. It used to wake you up, now it puts you to sleep.[1]

Malcolm's speeches swept us up, his voice resonant, warm, making the room disappear from consciousness. His love for his audience was palpable. At the time, I didn't know how to love myself. So when Malcolm spoke of how the violence of White society could resonate in Black self-hate, I understood this through my own self-hate. And as Malcolm taught his audience to transform their self-denigration into powerful, liberating self-love – through Black nationalism, Black speech, Black self-defence against White violence, and Black Islam in answer to White racist Christianity – he taught me about self-love too.

Like Richard Wright, Malcolm X had a voracious appetite for learning, reading, thinking. He combined this with an agile, probing intelligence, a fiery ethos and a comprehensive religious vision, first as a Black Muslim, then within a global Sunni Islam. All this took shape within the ghettos of America's most vibrant cities – New York, Chicago, Detroit, Los Angeles – the urban landscapes into which Northern Black life had been locked, the better to be dispossessed by Whites. Malcolm X took this landscape of urban apartheid and converted it into one of revolutionary spiritual, social, and economic transformation. His most important lesson of all was resistance by any means necessary to that which would exploit, denigrate and silence, whether that was American racial terror, the police state or his own (former) community in the Nation of Islam – a lesson for which he paid with his life.

What we students learned was profound: how words are housed in bodies, bodies in communities, communities in landscapes, and landscapes in the world at large, connecting past and present. Such embodied words have the power to create meaning imbued with enough force to create change across time and space, within ourselves, within the world. This was History, was Politics, was Literature, was Art, was Life itself.

I had been turning the assignment around in my mind's eye, probing it this way and that, terribly unsure. The night before it was due, I sat down and wrote it in one rush through the night into the morning. I chose the words that were the most potent and confusing in my life until that point, words I'd been avoiding.

Some years earlier, just after turning fifteen, I became friends with a neighbour who was nine years older. I was very lonely, living in a bedroom community outside New York City, going to a new school, uprooted and plunked down in a strange town where I knew no one, had no ties, was different from my schoolmates in ways I couldn't articulate but that all could feel so surely, as children do. My parents were busy elsewhere, in the city all day. Any attention, then, was a relief. The moment my neighbour came home from work, I would leave the emptiness of my house to ring her doorbell.

One night, she came out to me as gay. This was a nerve-wracking affair. We were in the midst of the AIDS epidemic and Cold War, Ronald Reagan at the helm, and the country caught up in a moral hysteria about homosexuality as an all-out assault on the natural order, social purity, and the security of America. Gays were despised and the news brought stories of gays killed, dragged behind a car, tied to barbed wire, beaten, stabbed.

I didn't care. I had never imagined myself as straight, had no name for myself except as a girl-who-was-a-boy. Gayness suited that.

After a while, she began to touch me. Soon after that I moved in with her, set up house, though still a child. I had not yet developed sexually. So at night I simply left my body, as necessary. I lent it out to her like a library book. This meant shutting down feeling, but that was better than loneliness. Moving into her house and social circle, a group of friends in their thirties, meant moving into a place in which I had a sort of place.

In some ways, I didn't mind lending out my body to her. I had hated my body from early on. It had betrayed me, with its femaleness. From the time I was very young I had thought, in the inchoate fashion of children, that if I hated myself enough, I might succeed in becoming what I yearned to be, a boy. So I discarded intimacy with this body. She could have it.

But the dissociation took its toll. After I left for college, I trusted no one. Anyone coming close, touching my skin, put me on high alert. As if seared by a branding iron's molten tip. I would jump away, run away even from those I let touch me.

And then I had a professor who asked us to write about words that made a difference. So I told him the truth.

About being in a bed in which I didn't want to be. Being touched as I shouldn't be. Mind splitting from body. Floating off somewhere, letting what had to happen, happen. And then, this business completed, to hear coming towards me the words 'I love you'. How those words would hang in the damp air, wanting an answer. Until my own words turned against me, I said what had to be said to keep my place – lent out my voice to be pimped, as I had my body.

What does it mean when we consent to our own abuse, our own silencing, our own erasure? And what if we already hated the body we were, the sex we were, divorcing and discarding it long before the abuser appeared?

It would take some years before I had good answers to this. Yet the effect of telling another this, telling myself this was shockingly electric. As the words took shape on the page the night I wrote them, an inner door opened behind which were locked thoughts, feelings, the physical and emotional kinetics of sexual abuse, social marginality, self-hate. Out came my voice. It was the first time that I spoke in public, to another as myself, in any real, meaningful sense. Once that happened, there was no going back.

The fact that this happened in the setting of a college course – and after this course in others – in which learning history, politics, and the arts was a practice that intimately conjoined emancipating self-discovery to rigorous intellectual, scholarly, and aesthetic endeavours profoundly shaped my understanding of scholarship in the broadest sense. To be a scholar was to fully engage the relation of learning to living, and the place of these in a world of beauty that at the same time was one of gross violence and injustice.

I could not have begun to speak if I had not studied Richard Wright, the young activists of SNCC and Malcolm X writing and speaking themselves into political, social, and literary being as something which did not yet exist: Black Americans whose humanity was fully affirmed, embraced, and safeguarded. In my courses, we studied their lives and words, and those of others, as intimate personal flows of sensibilities, politics, and aesthetics, of pain and beauty, of loss and achievement. By encountering them in this deeply personal fashion, we could learn from them how to engage the ways their landscapes and our own took shape through histories of inhumanity others demanded we forget. We learned not to look away from these, but to travel through them, to a future we would have a say in making. Following these, I learned too how to create a pathway for myself as queerly different and unhoused.

Academically, that pathway corresponds to the disciplines at whose intersection I've made an intellectual home for myself – politics, anthropology, and history – and to the core question at the heart of my scholarship: how to transform our planetary society from one in which large portions of humanity are made worthless and discardable into one in which each human life and each community is tangibly, visibly equal to every other?

Most fundamentally of all, however, I am still completing that first assignment, even now. To write, while feeling how the words of others have made all the difference, in my own life, but even more in the life that is ours together, as we seek to end the violence we do to one another, sometimes of the most brutal, comprehensive sort, without giving up the aspiration of hearing the other's voice break through the silence, also when that other voice is our own.

Suggestions for further reading

M. Marable. *Malcolm X: A Life of Reinvention*. New York, NY: Routledge, 2011.

D. McAdam. *Freedom Summer*. Oxford: Oxford University Press, 1988.

R. Wright. Ed. J.E. Wideman. *Black Boy*. New York, NY: Harper Perennial Modern Classics, 2020.

List of contributors

AGNES ANDEWEG obtained a PhD in Literature from Maastricht University. Her research interests encompass Dutch and Anglophone literature, gothic fiction, and Dutch cultural history and memory. At UCU, she teaches literature and research methodologies of the humanities.

ANNEMIEKE MEIJER studied English language & literature and obtained a PhD in eighteenth-century studies from Utrecht University. She worked as a higher education advisor in the field of internationalization and is currently UCU's head tutor and the coordinator of the Writing Centre.

ANTON E.M. VAN DE VEN obtained his PhD in theoretical physics from the State University of New York at Stony Brook. He conducted research and taught at universities in the USA, Germany, and the Netherlands. At UCU, he teaches physics and sustainability.

ALEXIS A. ARONOWITZ received her PhD in criminal justice from the University at Albany, N.Y. She has taught at universities in the United States, Germany, and the Netherlands and has worked as a consultant on human trafficking for numerous international organizations. Alexis has published extensively on the topic.

BALD DE VRIES studied law at Leiden University and obtained his PhD at Dublin City University, Ireland. Since 2001 he has taught legal theory at Utrecht University and has pioneered in educational innovation. He publishes in the field of legal and social theory as well as on academic legal education.

CHIARA ROBBIANO (PhD Classics, Leiden University; visiting professor Tohoku University, Japan) is philosophy lecturer and Honours Director at UCU. She publishes on cross-cultural philosophy in, among others, *Philosophy East and West* and *Ancient Philosophy*. She is involved in projects promoting dialogue (articles on intellectual virtues, TV series *Food For Thought*).

FILIPE FREIRE studied physics in Lisbon and received his PhD in particle physics and cosmology at Imperial College, London. He held post-docs in Germany, Ireland, and Leiden, before joining UCU, where he is now the physics track coordinator. He introduced the HiSPARC outreach project to measure cosmic rays to UCU.

FLORIS VAN DER BURG studied Liberal Arts at Bates College, philosophy at Leiden University, and obtained a PhD in philosophy from the University of Warwick. He taught philosophy at Warwick, the University of Amsterdam, and University College Utrecht, where he is Fellow in philosophy. He is the author of *Davidson and Spinoza*.

GAETANO FIORIN studied theoretical linguistics at the University of Verona and obtained a PhD in experimental linguistics from Utrecht University. He is a researcher affiliated to the Utrecht Institute of Linguistics. At UCU, he teaches logic and linguistics.

GERARD VAN DER REE teaches world politics, social innovation, and community engaged learning at UCU. His current field of research and action is the politics of the Anthropocene.

ROZI TÓTH is junior researcher of the Future of Academia project at Radboud University. She designs and researches innovative learning environments that contribute to enhancing students' agency in the climate crisis. She taught the courses International Relations and Creating Societal Impact at University College Utrecht.

GUUS DE KROM studied phonetics at Utrecht University and obtained his PhD on the acoustics and perception of pathological voice quality at the UiL-OTS Research Institute for Language and Speech. He worked as tutor and lecturer at university college, and taught courses in academic skills, research methods, and (bio)statistics.

JAMES KENNEDY has a doctorate in history from the University of Iowa. Until recently dean of University College Utrecht, he now teaches history and does work in community engaged learning at Utrecht University. A specialist of the modern Netherlands, he chaired the commission revising the Dutch canon.

JOCELYN BALLANTYNE earned her PhD in linguistics at the University of Texas at Austin. Two decades teaching in higher education has given her a deep appreciation of the value of exchanging knowledge between disciplines. At UCU, she teaches courses in linguistics, logic, and academic research skills.

KATJA RAKOW obtained her PhD in religious studies from Heidelberg University, Germany, and teaches Religious Studies at Utrecht University and UCU. Her research focuses on megachurches in the US and Singapore with a special interest in material religion, technology, and the use of texts in religious practices.

CORNELUS SANDERS trained at Nijmegen University and specialized at Maastricht University. He is a dermatologist who worked as an HIV and STI specialist in Zimbabwe and now works at the University Medical Centre Utrecht as a consultant in general dermatology and is a fellow of medical science at UCU.

MARKHA VALENTA researches and teaches at the intersection of politics, anthropology, and history. Her work focuses on public religion, migrant citizenship, the globalization of diversity politics, and community engagement. She is currently the Urban Citizen Fellow at the Netherlands Institute for Advanced Studies (NIAS).

MARY BOUQUET holds a PhD in Anthropology from Cambridge University and taught and conducted research at universities and museums in Portugal, Norway, and the Netherlands, where she also curated exhibitions. She is co-editor of the Berghahn Books Museums and Collections Series. At UCU, she teaches anthropology and museum studies.

ROBERT DUNN is a clinical psychologist and holds a PhD from the Professional School of Psychology in San Francisco. He has taught at University College Utrecht since 2001, is currently retired, but still coordinates the UCU Global Mental Health Project.

TIJANA ŽAKULA has a PhD in art history from Utrecht University and is currently working as a lecturer and researcher at UCU. Her specialty is Dutch seventeenth-century art with a particular emphasis on painting and art theory. She contributed to the collection catalogue of the Rijksmuseum and authored a book and a number of articles on Early Modern art and art theory.

VIKTOR BLÅSJÖ has a PhD in the history of mathematics in the seventeenth century. He has also taught at Marlboro College, a liberal arts college in Vermont. You can follow him on Twitter @viktorblasjo and listen to his Opinionated History of Mathematics podcast.

List of figures

Notes

Floris van der Burg: The indispensable Truth

1. H. Ahrendt. *The Origins of Totalitarianism* (Meridian Books, 1958), 474.
2. S.T. Coleridge. *Biographia Literaria*. London: Rest Fenner, 1817, ch. 14.
3. J. Hughes. *Ferris Bueller's Day Off*. Paramount Pictures, 1986.
4. C. Butler. *Postmodernism: A Very Short Introduction*. Oxford: Oxford University Press, 2002, ch. 2f.

Agnes Andeweg: Fictionality, or the importance of being earnest

1. K. Lodge. 'The Use of the Past Tense in Games of Pretend'. *Journal of Child Language*, 6(2) (1979), 365-69. https://doi:10.1017/S030500090000236

Alexis A. Aronowitz: And justice for all

1. D.L. Myers. 'Juvenile Transfer to Adult Court: Ongoing Search for Scientific Support.' *Criminology & Public Policy*, 15 (2016), 927-938. doi:10.1111/1745-9133.12232.
2. R.V. Clarke, M. Felson. *Criminology, Routine Activity, and Rational Choice*. New Brunswick, NJ: Transaction Publishers, 1993.
3. C. Beccaria. *On Crimes and Punishments*. New York, NY: Bobbs-Merrill, 1963. https://pdfs.semanticscholar.org/f69e/c38c8cbb4f5a8b2febao91ebeaa7b-808f7d6.pdf
4. S.N. Zane, B.C. Welsh, D.P. Mears. 'Juvenile Transfer and the Specific Deterrence Hypothesis: Systematic Review and Meta-analysis.' *Criminology & Public Policy*, 15 (2016), 901-925. https://diginole.lib.fsu.edu/islandora/object/fsu:640071/datastream/PDF/view
5. C.R. Bartol, A.M. Bartol. *Criminal Behavior: A Psychological Approach*. Englewood Cliffs, NJ: Prentice Hall, 2017.
6. T.P. Thornberry, M.D. Krohn, Eds. *Taking Stock of Delinquency: An Overview of Findings from Contemporary Longitudinal Studies*. Amsterdam: Kluwer Academic Publishers, 2004; S. Krutikova, H.B. Lilleør. 'Fetal Origins of Personality: Effects of Early Life Circumstances on Adult Personality Traits'. *Psychology*, 2015. https://editorialexpress.com/cgibin/conference/download.cgi?db_name=NEUDC2015&paper_id=226.
7. C.R. Bartol, A.M. Bartol. *Criminal Behavior: A Psychological Approach*. Englewood Cliffs, NJ: Prentice Hall, 2017.
8. L. Murray, P.J. Cooper. 'Postpartum depression and child development'. *Psychological Medicine*, 27(2) (March 1997), 253-260. doi: https://doi.org/10.1017/S0033291796004564.
9. I. Bretherton. 'The Origins of Attachment Theory: John Bowlby and Mary Ainsworth'. *Developmental Psychology*, 28(5) (1992),759-775; T. Hirschi. *Causes of Delinquency*. Berkeley, CA: University of California Press, 1969.

10. M. Hoeve, J.M. Stams, C.E. van der Put, J. Semon Dubas, P.H. van der Laan, J.R.M. Gerris. 'A Meta-analysis of Attachment to Parents and Delinquency'. *Journal of Abnormal Child Psychology*, 40 (2012), 771. https://doi.org/10.1007/s10802-011-9608-1

11. M. Hoeve, J.M. Stams, C.E. van der Put, J. Semon Dubas, P.H. van der Laan, J.R.M. Gerris. 'A Meta-analysis of Attachment to Parents and Delinquency'. *Journal of Abnormal Child Psychology*, 40 (2012); T. Hirschi. *Causes of Delinquency*. Berkeley, CA: University of California Press, 1969.

12. C.R. Bartol, A.M. Bartol. *Criminal Behavior: A Psychological Approach*. Englewood Cliffs, NJ: Prentice Hall, 2017.

13. J.E. Lansford, S. Miller-Johnson, L.J. Berlin, K. Dodge, J.E. Bates, S. Gregory. 'Early Physical Abuse and Later Violent Delinquency: A Prospective Longitudinal Study'. *Child Maltreatment*, 12(3) (August 2007), 233-245. https://www.ncbi.nlm.nih.gov/pmc/articles/PMC2771618/

14. C.R. Bartol, A.M. Bartol. *Criminal Behavior: A Psychological Approach*. Englewood Cliffs, NJ: Prentice Hall, 2017.

15. C.R. Bartol, A.M. Bartol. *Criminal Behavior: A Psychological Approach*. Englewood Cliffs, NJ: Prentice Hall, 2017.

16. D. Finkelhor, R.K. Ormrod, & H.A. Turner (2007a), cited in: J.D. Ford, B.C. Delker. 'Polyvictimization in Childhood and its Adverse Impacts Across the Lifespan: Introduction to the Special Issue'. *Journal of Trauma & Dissociation*, 19(3) (2018), 275-288. doi: 10.1080/15299732.2018.1440479

17. J.D. Ford, B.C. Delker. 'Polyvictimization in Childhood and its Adverse Impacts Across the Lifespan: Introduction to the Special Issue'. *Journal of Trauma & Dissociation*, 19(3) (2018), 275-288.

18. H.C. Hahm, Y. Lee, A. Ozonoff, M.J. van Wert. 'The Impact of Multiple Types of Child Maltreatment on Subsequent Risk Behaviors Among Women During the Transition from Adolescence to Young Adulthood'. *Journal of Youth Adolescence*, 39(5) (2010), 528-540. doi:10.1007/s10964-009-9490-0

19. M. Dong, R.F. Anda, V.J. Felitti, S.R. Dube, D.F. Williamson, T.J. Thompson, W.H. Giles. 'The Interrelatedness of Multiple Forms of Childhood Abuse, Neglect, and Household Dysfunction'. *Child Abuse & Neglect*, 28(7) (2004), 771-784.

20. R.M. Post 1992; Z.V. Segal, J.M. Williams, J.D. Tesadale, & M. Gemar M. (1996) cited in: H.C. Hahm, Y. Lee, A. Ozonoff, M.J. van Wert. 'The Impact of Multiple Types of Child Maltreatment on Subsequent Risk Behaviors Among Women During the Transition from Adolescence to Young Adulthood'. *Journal of Youth Adolescence*, 39(5) (2010), 528-540.

21. K.A. Dodge, M.T. Greenberg, P.S. Malone. 'Testing an Idealized Dynamic Cascade Model of the Development of Serious Violence in Adolescence'. *Child Development*, 79(6) (2008), 1907-1927. doi:10.1111/j.1467-8624.2008.01233.x; https://www.ncbi.nlm.nih.gov/pmc/articles/PMC2597335/

22. Federal Bureau of Investigation. 'Ten-Year Arrest Trends; By Sex'. *Crime Trends in the United States: Uniform Crime Report*, Table 33 (2020). https://

ucr.fbi.gov/crime-in-the-u.s/2019/crime-in-the-u.s.-2019/topic-pages/tables/table-33y Sex, 2010-2019

23.   R.N. Parker, K. Auerhahn. 'Alcohol, Drugs, and Violence'. *Annual Review of Sociology*, 24 (1998), 291-311.

24.   S. Green, A. Pemberton. 'The Impact of Crime: Victimization, Harm and Resilience'. S. Walklate, Ed., *Handbook of Victims and Victimology*. Milton Park: Taylor & Francis, 2017, 77-102. https://doi.org/10.4324/9781315712871-6

25.   P. Brennan, A. Raine. 'Biological Bases of Anti-Social Behavior: Psycho-physiological, Neurological and Cognitive Factors'. *Clinical Psychology Review*, 17(6) (1997), 589-604.

26.   M. Wolfgang, R. Figlio, T. Selin. *Delinquency in a Birth Cohort.* Chicago, IL: University of Chicago Press, 1972.

Jocelyn Ballantyne: Handling tricky questions

1.   H. Cooper. 'Trump team's queries about Africa point to skepticism about aid'. *New York Times*, 13 January 2017. https://www.nytimes.com/2017/01/13/world/africa/africa-donald-trump.html

2.   L. Karttunen. 'Syntax and Semantics of Questions'. *Linguistics and Philosophy*, 1(1) (1977), 3-44.

3.   J. Searle. *Speech Acts.* Cambridge: Cambridge University Press, 1969.

4.   H.P. Grice. 'Presupposition and Conversational Implicature'. P. Cole, Ed., *Radical Pragmatics*. New York, NY: Academic Press, (1981), 183-198.

5.   R. Stalnaker. 'Pragmatics'. D. Davidson, G. Harman, Eds., *Semantics of Natural Language*. Dordrecht: Reidel, (1972), 389-408.

Gaetano Fiorin: What is meaning?

1.   L. Wittgenstein, G.E.M. Anscombe. *Philosophical Investigations*. London: Basic Blackwell, 1953.

2.   W.V.O. Quine. *Word and Object.* Cambridge, MA: MIT Press, 1960.

3.   H. Putnam. 'The Meaning of "Meaning"'. *Minnesota Studies in the Philosophy of Science*, 7 (1975), 131-193.

4.   N. Chomsky. *New Horizons in the Study of Language and Mind.* Cambridge: Cambridge University Press, 2000.

5.   F. de Saussure. *Cours de linguistique générale, publié par Ch. Bally et A. Sechehaye avec la collaboration de A. Riedlinger.* Paris: Payot, 1916.

6.   T. O'Connor. 'Emergent Properties'. Ed. E.N. Zalta. *The Stanford Encyclopedia of Philosophy* (online Fall 2020 Edition). https://plato.stanford.edu/archives/fall2020/entries/properties-emergent/

7.   C.K. Ogden, I.A. Richards. *The Meaning of Meaning.* London: Kegan Paul, 1923.

8.   G. Fiorin, D. Delfitto. *Beyond Meaning: A Journey across Language, Perception and Experience.* Dordrecht: Springer, 2020.

Chiara Robbiano: Parmenides and Dōgen – an encounter

1.   D. Loy. *Nonduality: A Study in Comparative Philosophy*. Amherst, NY: Pro-
      metheus Books, 2012.

2.   C. Robbiano. 'Being is Not an Object: An Interpretation of Parmenides'
      Fragment DK B2 and a Reflection on Assumptions'. *Ancient Philosophy*, 36
      (2016), 263-301.

3.   C. Robbiano. 'Can Words Carve a Jointless Reality? Parmenides and
      Sankara'. *Journal of World Philosophies*, 3:1 (2018), 31-43.https://scholarworks.
      iu.edu/iupjournals/index.php/jwp/article/view/1615

4.   E. Deutsch. *Advaita Vedānta: A Philosophical Reconstruction*. Honolulu:
      University of Hawaii Press, 1968.

5.   T.P. Kasulis. *Zen action/Zen person*. Honolulu: University of Hawaii Press,
      1981; B.W. Davis. 'The Philosophy of Zen Master Dōgen: Egoless Perspectiv-
      ism'. *The Oxford Handbook of World Philosophy*. Oxford: Oxford University
      Press, 2011, 348-60.

6.   G. Kopf. 'Antiphony: A Model of Dialogue'. *The Bulletin of the Nanzan Institute
      for Religion and Culture*, 39 (2015), 26-35. https://nirc.nanzan-u.ac.jp/nfile/4393;
      G. Kopf. "Peace through Self-Awareness: A Model of Peace Education Based on
      Buddhist Principles". *Journal of Ecumenical Studies*, 50:1 (2015), 47-56.

7.   H. Diels, W. Kranz. *Die Fragmente der Vorsokratiker*. Berlin: Weidmannn,
      1996 [cited as 'DK' (Diels Kranz) plus B (quotation from Parmenides) fol-
      lowed by the quotation ("fragment") number].

8.   Eds. G. Nishijima, C. Cross. *Dōgen, Shōbōgenzō*. Charleston, NC: Book Surge
      Publishing, vols. 1-4, 1994; Ed. K. Tanahashi. *Moon in a Dewdrop: Writings of
      Zen Master Dogen*. London: Macmillan, 1985; Dōgen, *Shōbōgenzō*. http://
      www.shomonji.or.jp/soroku/genzou.htm.

9.   Eds. G. Colli, M. Montanari. F. Nietzsche. *Sämtliche Werken: Kritische Studi-
      enausgabe in 15 Bänden. Band 1, Die Philosophie im tragischen Zeitalter der
      Griechen*. Munich: De Gruyter, 1980, 844.

Cornelus Sanders: On being a doctor

1.   A. Kleinman. 'Presence'. *Lancet*, 389 (2017), 2466-2467.

2.   T.J. Kaptchuk, F.G. Miller. 'Placebo Effects in Medicine'. *New England Journal
      of Medicine*, 373 (2015), 8-9. doi: 10.1056/NEJMp1504023

3.   L. Colloca, A.J. Barsky. 'Placebo and Nocebo Effects'. *New England Journal of
      Medicine*, 382 (2020), 554-61. doi: 10.1056/NEJMra1907805

4.   J. Zaki J. 'The Caregiver's Dilemma: In Search of Sustainable Medical Empa-
      thy'. *Lancet*, 396 (2020), 458-459.

5.   E.J. Emanuel, E. Gudbranson. 'Does Medicine Overemphasize IQ?' *JAMA*,
      319 (2018), 651-52. doi: 10.1001/jama.2017.20141

6.   B. Lyons, M. Gibson, L. Dolezal. 'Stories of Shame'. *Lancet*, 391 (2018). 1568-1569.

7.   A. Verghese, N.H. Shah, R.A. Harrington. 'What This Computer Needs Is a
      Physician: Humanism and Artificial Intelligence'. *JAMA*, 319 (2018), 19-20.
      doi: 10.1001/jama.2017.19198

Bald de Vries: Law, imagination, and poetry

1.   S.G. Stolberg. 'Emotional Schiff Speech Goes Viral. Delighting the Left and
      Engaging the Right'. *The New York Times*, 24 January 2020. https://www.
      nytimes.com/2020/01/24/us/politics/adam-schiff-closing-remarks.html

2.   T. Hobbes. *Leviathan*. Harmondsworth: Penguin Books, 2017, ch. xiii.

3.   J.B. White. 'Doctrine in a Vacuum: Reflections on What a Law School Ought
      (and Ought Not to Be)'. *Journal of Legal Education*, 36 (1986), 155-166.

4.   N. Luhmann. *Law as a Social System*. Oxford: Oxford University Press, 2004.

5.   P. Scholten, Ed. *Mr. C. Asser's handleiding tot de beoefening van het Neder-
      lands Burgerlijk recht: Algemeen Deel*. Zwolle: Tjeenk Willink, 1931. https://
      www.paulscholten.eu/

6.   I. van Domselaar. 'Moral Quality in Adjudication: On Judicial Virtues and
      Civic Friendship'. *Netherlands Journal of Legal Philosophy*, 44(1) (2015), 24-
      46.

7.   P.B. Shelley. 'A Defence of Poetry'. *Selected Poems and Prose*. Harmonds-
      worth: Penguin Books, 2017, 635.

8.   P.B. Shelley. 'A Defence of Poetry'. *Selected Poems and Prose*. Harmonds-
      worth: Penguin Books, 2017, 635-665.

9.   P.B. Shelley. 'A Defence of Poetry'. *Selected Poems and Prose*. Harmonds-
      worth: Penguin Books, 2017, 635-665.

10.   R.M. Hutchins, M.J. Adler. 'Gateway to the great books'. Toronto: Encyclo-
      paedia Britannica, 1963, 5: 241.

11.   P.B. Shelley. 'A Defence of Poetry'. *Selected Poems and Prose*. Harmonds-
      worth: Penguin Books, 2017, 660, 640.

12.   P.B. Shelley. 'A Defence of Poetry'. *Selected Poems and Prose*. Harmonds-
      worth: Penguin Books, 2017, 557, 560.

13.   P.B. Shelley. 'A Defence of Poetry'. *Selected Poems and Prose*. Harmonds-
      worth: Penguin Books, 2017, 642.

14.   P.B. Shelley. 'A Defence of Poetry'. *Selected Poems and Prose*. Harmonds-
      worth: Penguin Books, 2017, 655.

15.   J. Gray. 'Hobbes, our great contemporary'. *The New Statesman*, 20 Septem-
      ber 2012.

16.   J.B. White. 'Intellectual Integration'. *Issues in Interdisciplinary Studies*, 5(1)
      (1987), 13.

17.   B. de Vries. 'Law, Imagination and Poetry: Using Poetry as a Means of Learn-
      ing'. *Law & Method*, January 2019. https://doi.org/10.5553/REM/.000039.

18.   M. Sandel. *Justice: What's the Right Thing to Do?* Harmondsworth: Penguin
      Books, 2009.

19.   M. Sandel. *Justice: What's the Right Thing to Do?* Harmondsworth: Penguin
      Books, 2009, 27-28.

20.   J. Frank. *Law and the Modern Mind*. City: Routledge, 2009, 109.

21.   B. van Klink, B. de Vries. 'Skeptical Legal Education'. *Law & Method*, 2013, 49.
      https://doi.org/10.5553/ReM/221225082013003002004

Katja Rakow: Religion 2.1

1. J. Fox. 'Secularization'. J. Hinnels, Ed., *The Routledge Companion to the Study of Religion* (second edition). Abingdon: Routledge, 2010, 308-312.
2. G. Davie. 'Mainstream Religion in the Western World'. *The Sociology of Religion*. London: Sage, 2007, 135-157.
3. L. Woodhead, C.H. Partridge, H. Kawanami, Eds. *Religions in the Modern World: Traditions and Transformations* (third edition). London: Routledge, 2016.
4. J. McCurry. 'Japan: Robot Dogs Get Solemn Buddhist Send-off at Funerals'. *The Guardian*, 3 May 2018. https://www.theguardian.com/world/2018/may/03/japan-robot-dogs-get-solemn-buddhist-send-off-at-funerals.
5. P. Holley. 'Innovations: Meet 'Mindar' the Robotic Buddhist priest'. *The Washington Post*, 22 August 2019. https://www.washingtonpost.com/technology/2019/08/22/introducing-mindar-robotic-priest-that-some-are-calling-frankenstein-monster/.
6. M.B. McGuire. *Lived Religion: Faith and Practice in Everyday Life*. Oxford: Oxford University Press, 2008.
7. C. Martin. *A Critical Introduction to the Study of Religion*. London: Routledge, 2017, 117.
8. P.G. Horsfield. *From Jesus to the Internet: A History of Christianity and Media*. Chichester: Wiley Blackwell, 2015, 193.
9. P.G. Horsfield. *From Jesus to the Internet: A History of Christianity and Media*. Chichester: Wiley Blackwell, 2015, 194.
10. P.G. Horsfield. *From Jesus to the Internet: A History of Christianity and Media*. Chichester: Wiley Blackwell, 2015, 195.
11. F. Furedi. *Power of Reading: From Socrates to Twitter*. London: Bloomsbury Continuum, 2015, 57-80.
12. K. Rakow. 'The Bible in the Digital Age: Negotiating the Limits of "Bibleness" of Different Bible Media'. M. Opas, A. Haapalainen, Eds., *Christianity and the Limits of Materiality*. London: Bloomsbury, 2017, 110.
13. B. Meyer. 'Medium'. S.B. Plate, Ed., *Key Terms in Material Religion*. London: Bloomsbury Academic, 2015, 141.
14. C. Martin. *A Critical Introduction to the Study of Religion*. London: Routledge, 2017, 120; Jeremy Stolow. 'Introduction: Religion, Technology, and the Things in Between'. *Deus in Machina: Religion, Technology, and the Things in Between*. New York, NY: Fordham University Press, 2012, 8. [Image: Print Bible and Bible App on smartphone, photo by the author]

Robert Dunn: Global mental health and the evolution of clinical psychology

1. D.R. Singla, B.A. Kohrt, L.K. Murray, A. Anand, B.F. Chorpita, V. Patel. 'Psychological Treatments for the World: Lessons from Low – and Middle – Income Countries'. *Annual Review of Clinical Psychology*, 13 (2017), 149-181.

2.   World Health Organization. *Mental Health Action Plan 2013-2020*. Geneva: WHO Press, 2013.

3.   Patientenfederatie Nederland. 'Wachtlijsten in de ggz'. https://kennisbank. patientenfederatie.nl/app/answers/detail/a_id/419/~/wachtlijsten-in-de-ggz, 2020.

4.   World Health Organization. *Mental Health Atlas 2011*. Geneva: WHO Press, 2011.

Rozi Tóth and Gerard van der Ree: Heroes of the in-between

1.   M. Heidegger. 'The Origin of the Work of Art'. *Basic Writings*. New York, NY: Harper Collins, 1993, 165-182.

2.   B.A. Bentley. *Scientific Cosmology and International Orders*. Cambridge: Cambridge University Press, 2018, 29-74.

3.   B. Latour. *Facing Gaia: Eight Lectures on the New Climatic Regime*. Cambridge: Polity Press, 2017, 41-74.

4.   M. Arias-Maldonado. *Environment and Society: Socionatural Relations in the Anthropocene*. London: Springer, 2015, 33-71.

5.   D. Rothe. 'Governing the End Times? Planet Politics and the Secular Eschatology of the Anthropocene'. *Millennium Journal of International Studies* (2019), 1-22.

6.   R. Scranton. *Learning to Die in the Anthropocene: Reflections on the End of a Civilization*. San Francisco, CA: City Lights, 2015, 17-21.

7.   B. Latour. *Down to Earth: Politics in the New Climatic Regime*. London: Polity Press, 2018.

8.   D. Haraway. *Staying with the Trouble: Making Kin in the Chthulucene*. Durham, NC: Duke University Press, 2016, 34.

9.   Roy Scranton. *Learning to Die in the Anthropocene: Reflections on the End of a Civilization*. San Francisco, CA: City Lights, 2015, 24.

10.   Donna Haraway. *Staying with the Trouble: Making Kin in the Chthulucene*. Durham, NC: Duke University Press, 2016, 38.

11.   J. Campbell. *The Hero with a Thousand Faces*. Novato, CA: New World Library, 2008, 23-31.

12.   Donna Haraway, *Staying with the Trouble: Making Kin in the Chthulucene*. Durham, NC: Duke University Press, 2016, 1-8.

13.   B. Latour. *Down to Earth: Politics in the New Climatic Regime*. London: Polity Press, 2018, 64.

14.   B. Brown. *Dare to Lead: Brave Work. Tough Conversations. Whole Hearts*. London: Vermillion, 2018, 11.

Viktor Blåsjö: What history's most overqualified calculus student tells us about Liberal Arts mathematics

1.   C.I. Gerhardt, Ed. *Leibnizens mathematische Schriften*. Halle: H.W. Schmidt, 1860, 2.2.370.

2. J.H. Hofmann. *Leibniz in Paris 1672-1676: His Growth to Mathematical Maturity*. Cambridge: Cambridge University Press, 1974.

3. *Œuvres complètes de Christiaan Huygens publiées par la Société Hollandaise des Sciences*. The Hague: Nijhoff, 1888-1950. Vol. 10, Letter 2822. https://dbnl.nl/tekst/huyg003oeuv00_01/.

4. *Œuvres complètes de Christiaan Huygens*. Vol. 10, Letter 2829.

5. *Œuvres complètes de Christiaan Huygens*. Vol. 10, Letter 2693.

6. *Œuvres complètes de Christiaan Huygens*. Vol. 10, Letter 2699.

7. *Œuvres complètes de Christiaan Huygens*. Vol. 9, Letter 2632.

8. *Œuvres complètes de Christiaan Huygens*. Vol. 10, Letter 2660.

9. F. Klein. 'Über Aufgabe und Methode des mathematischen Unterrichts an den Universitäten'. *Jahresbericht der Deutschen Mathematiker-Vereinigung*, 7 (1899), 133.

10. Plato. *Republic*, VII.537c.

11. F. Klein. *Vorlesungen über die Entwicklung der Mathematik im 19. Jahrhundert*, Teil I. Berlin: Springer-Verlag, 1928; quoted from the English translation: *Development of Mathematics in the 19th Century*. Brookline, MA: Mathematical Science Press, 1979, 152.

Guus de Krom: Statistics

1. W.A. Guy. 'On the Original and Acquired Meaning of the Term "Statistics", and on the Proper Functions of a Statistical Society: also on the Question Whether There be a Science of Statistics; and, if so, What are its Nature and "Social Science"'. *Journal of the Statistical Society of London*, 28(4) (1865), 478-493.

2. S.M. Stigler. *The History of Statistics: The Measurement of Uncertainty Before 1900*. Cambridge, MA: Harvard University Press, 1986.

3. S.M. Stigler. *The History of Statistics: The Measurement of Uncertainty Before 1900*. Cambridge, MA: Harvard University Press, 1986.

4. L.A.J. Quetelet. *Sur l'homme et le développement de ses facultés, ou Essai de physique sociale*, vol. 2. Paris: Bachelier, 1869.

5. S.J. Gould. *The Mismeasure of Man* (revised edition). New York: W.W. Norton & Co, 1996.

6. D.S. Moore. 'Statistics among the Liberal Arts'. *Journal of the American Statistical Association*, 93(444) (1998), 1253-1259. doi:10.1080/01621459.1998.10473786

7. J. Best. 'Lies, Calculations and Constructions: Beyond How to Lie with Statistics'. *Statististical Science*, 20(3) (2005), 210-214. doi:10.1214/088342305000000232

8. J. Best. 'Lies, Calculations and Constructions: Beyond How to Lie with Statistics'. *Statististical Science*, 20(3) (2005), 210-214. doi:10.1214/088342305000000232

9. M.L. Ambrose. 'Lessons from the Avalanche of Numbers: Big Data in Historical Perspective'. *ISJLP*, 11 (2015), 201-277.

10. M.L. Ambrose. 'Lessons from the Avalanche of Numbers: Big Data in Historical Perspective'. *ISJLP*, 11 (2015), 201-277.

11. J. Best. 'Lies, Calculations and Constructions: Beyond How to Lie with Statistics'. *Statististical Science*, 20(3) (2005), 210-214. doi:10.1214/088342305000000232

12. S.J. Gould. *The Mismeasure of Man* (revised edition). New York: W.W. Norton & Co, 1996.

13. R. Creemers. 'China's Social Credit System: An Evolving Practice of Control', 2018. Available at SSRN: https://ssrn.com/abstract=3175792 or http://dx.doi.org/10.2139/ssrn.31757922

14. J. Dressel, H. Farid. 'The Accuracy, Fairness, and Limits of Predicting Recidivism'. *Science Advances*, 4(1) (2018). doi: 10.1126/sciadv.eaao5580.

15. J. Dressel, H. Farid. 'The Accuracy, Fairness, and Limits of Predicting Recidivism'. *Science Advances*, 4(1) (2018). doi: 10.1126/sciadv.eaao5580.

16. D.S. Moore. 'Statistics among the Liberal Arts'. *Journal of the American Statistical Association*, 93(444) (1998), 1253-1259. doi:10.1080/01621459.1998.10473786

17. A. Tversky, D. Kahneman. 'Belief in the Law of Small Numbers'. *Psychological Bulletin*, 76(2) (1971), 105.

18. M.H. van Rijn, A. Bech, J. Bouyer, J.A. van den Brand. 'Statistical Significance versus Clinical Relevance'. *Nephrology Dialysis Transplantation*, 32, supplement 2 (2017), ii6-ii12.

19. M.L. Head, L. Holman, R. Lanfear, A.T. Kahn, M.D. Jennions. 'The Extent and Consequences of P-hacking in Science'. *PLoS Biology*, 13(3) (2015). doi:http://doi.org/10.1371/journal.pbio.1002106

20. J.P.A. Ioannidis. 'Why Most Published Research Findings Are False'. *PLoS Medicine*, 2(8) (2005), 696-701.

21. J.J. Arnett. 'The Neglected 95%: Why American Psychology Needs to Become Less American'. *American Psychologist*, 63(7) (2008), 602-614.

22. N. Rose. 'Governing by Numbers: Figuring Out Democracy'. *Accounting, Organizations and Society*, 16(7) (1991), 673-692. doi:https://doi.org/10.1016/0361-3682(91)90019-B

James Kennedy: The canon of the Netherlands revisited

1. Ed. F.P. van Oostrom. *entoen.nu. De Canon van Nederland. Rapport van de Commissie Ontwikkeling Nederlandse Canon, Deel A*. Den Haag: Ministerie van OCW, 2006, 15-23.

2. Commissie Herijking Canon van Nederland. *Open vensters voor onze tijd. De Canon van Nederland herijkt*. Amsterdam: Amsterdam University Press, 2020.

3. N. Sleiderink. 'Vlaamse canon ziet over twee jaar levenslicht'. *De Tijd*, 18 September 2020.

4. C. Haas. 'The History-Canon Project as Politics of Identity: Renationalizing History Education in Denmark'. *History Education Research Journal*, 15(2) (2019), 180-192.

5. Commissie Herijking Canon van Nederland. *Open vensters voor onze tijd.*
    De Canon van Nederland herijkt. Amsterdam: Amsterdam University Press,
    2020, 14-15.

6. C. Haas. 'The History-Canon Project as Politics of Identity: Renationalizing
    History Education in Denmark'. *History Education Research Journal,* 15(2)
    (2019), 180-192.

7. M. Kropman. 'elkaar interviewen'. M. Lebouille, B. Ros, Eds. '*Verder... na
    Parijs'. Didactief,* 50(9) (November 2020).

8. P. Emmer. 'Sla maar over'. *Kleio,* 61(6) (2020), 66-68; K. Kleijn. 'Niets in te
    brengen: Maria van Bourgondië in de Canon'. *De Groene Amsterdammer,* 28-
    29 (2020); J. Kennedy. 'Aanzet tot een meerstemmige discussie: Reactie van
    de canoncommissie op Piet Emmer'. *Kleio,* 61(6) (2020), 70-71; K. Goudsmit,
    J. Kennedy, H. Tuithof. 'Nabeschouwing op de Herijking Canon van Neder-
    land'. *Tijdschrift voor Geschiedenis,* 2021.

9. Ed. F.P. van Oostrom. *entoen.nu. De Canon van Nederland. Rapport van de
    Commissie Ontwikkeling Nederlandse Canon, Deel A.* Den Haag: Ministerie
    van OCW, 2006, 23-24.

10. K. Kleijn. 'Niets in te brengen: Maria van Bourgondië in de Canon'. *De
    Groene Amsterdammer,* 28-29 (2020).

11. M. Doorman. *Boekman 125: 50 jaar Internationaal Cultuurbeleid* (December
    2020), 56-57.

Tijana Žakula: Love thyself

1. Ovid. *Metamorphoses.* Ed. E.J. Kenney, trans. A.D. Melville. Oxford: Oxford
    University Press, 1986, 61-66.

2. L-B. Alberti. *On Painting and on Sculpture: The Latin texts of De pictura and
    De statua.* Ed. C. Grayson. London: Phaidon Press, 1972, 98-100.

3. Pliny. *Natural History.* Trans. H. Rackham, 10 vols. Cambridge, MA: Harvard
    University Press, 1959, 9: 327, 329, 263.

4. É. Pommier. *Théories du portrait: de la Renaissance aux Lumières.* Paris: Gal-
    limard, 1998, 120-21.

5. Pliny. *Natural History.* Trans. H. Rackham, 10 vols. Cambridge, MA: Harvard
    University Press, 1959, 9: 373.

6. Ed. J. Woodall. *Portraiture: Facing the Subject.* Manchester: Manchester
    University Press, 1997, 77.

7. É. Pommier. *Théories du portrait: de la Renaissance aux Lumières.* Paris: Gal-
    limard, 1998, 223.

8. J. Fletcher. 'The Renaissance Portrait: Function,Uses and Display'. L. Camp-
    bell et. al. *Renaissance Faces: van Eyck to Titian,* London: National Gallery,
    2008, 53; É. Pommier. *Théories du portrait: de la Renaissance aux Lumières.*
    Paris: Gallimard, 1998, 128-29.

9. G.P. Lomazzo. *Trattato dell'arte de la pittura.* Milan: Paolo Gottardo Pontio,
    1584, 432.

10. G. de Lairesse. *Groot schilderboek.* Amsterdam: Hendrick Desbordes, 1712, 2:7.
11. G. de Lairesse. *Groot schilderboek.* Amsterdam: Hendrick Desbordes, 1712, 2:7-8.
12. G. de Lairesse. *Groot schilderboek.* Amsterdam: Hendrick Desbordes, 1712, 2:8.
13. G. de Lairesse. *Groot schilderboek.* Amsterdam: Hendrick Desbordes, 1712, 2:30.
14. G. de Lairesse. *Groot schilderboek.* Amsterdam: Hendrick Desbordes, 1712, 2:31.
15. L. Campbell. *Renaissance Portraits: European Portrait-painting in the 14th, 15th and 16th Centuries.* New Haven, CT: Yale University Press, 1990, 124.
16. E.S. Gordenker. *Anthony van Dyck (1599-1641) and the Representation of Dress in Seventeenth-century Portraiture.* Turnhout: Brepols, 2001, 73.
17. A. Ribeiro. *Ingres in Fashion: Representations of Dress and Appearance in Ingres's Images of Women.* New Haven, CT: Yale University Press, 1999.
18. G. Tinterow, P. Conisbee, H. Naef. *Portraits by Ingres: Image of an Epoch.* New York, NY: Harry N. Abrams, 1999, 418.
19. G. Tinterow, P. Conisbee, H. Naef. *Portraits by Ingres: Image of an Epoch.* New York, NY: Harry N. Abrams, 1999, 416.
20. T. Žakula. *Reforming Dutch Art: Gerard de Lairesse on Beauty, Morals and Class.* Amsterdam: Stichting voor Nederlandse Kunsthistorische Publicaties, 2015, 97.
21. S. Holiday, M.J. Lewis, R. Nielsen, H.D. Anderson, M. Elinzano. 'The Selfie Study: Archetypes and Motivations in Modern Self-Photography'. *Visual Communications Quarterly* 23(3) (2016), 175-87.

Mary Bouquet: Two monuments

1. R.K. Younger. 'Making Memories, Making Monuments: Changing Understandings of Henges in Prehistory and the Present'. Ed. K. Brophy, *The Neolithic in Mainland Scotland.* Edinburgh: Edinburgh University Press, 2016, 116-38.
2. Public Art/ Architecture/ Sculpture/ Painting and Decorative Designs for the General Public. http://www.visual-arts-cork.com/public-art.htm.
3. F, Choay. *The Invention of the Historic Monument.* Cambridge: Cambridge University Press, 2001.
4. Het Andere Monument/ Home. https://www.hetanderemonument.nl/
5. M. Mauss. *The Gift: The Form and Reason for Exchange in Archaic Societies.* London: Routledge, 1990.
6. A. Gell. *Art and Agency: An Anthropological Theory.* Oxford: Oxford University Press, 1998; R. Sansi. *Art, Anthropology and the Gift.* London: Bloomsbury, 2015.
7. N. van der Voort, F. Vanclay. 'Social Impacts of Earthquakes caused by Gas Extraction in the Province of Groningen, The Netherlands'. *Environmental Impact Assessment Review,* 50 (2015), 2.
8. Het Andere Monument/ Home. https://www.hetanderemonument.nl/
9. N. van der Voort, F. Vanclay. 'Social Impacts of Earthquakes caused by Gas Extraction in the Province of Groningen, The Netherlands'. *Environmental Impact Assessment Review,* 50 (2015), 3.

10. N. van der Voort, F. Vanclay. 'Social Impacts of Earthquakes caused by Gas Extraction in the Province of Groningen, The Netherlands'. *Environmental Impact Assessment Review,* 50 (2015), 6.

11. Het Slochter Molecule. https://dhaps.org/kunstwerken/het-slochter-molecule/

12. Atomium/discover/The Origins. https://www.Atomium.be

13. J.J. Wiersma. *Bakstain. Een onderzoek naar de baksteenfabricage op het Hogeland en in de Noordoost-Fivelingo en het Centrale Woldgebied (provincie Groningen) in de periode circa 1550-1960, met speciale aandacht voor de steenbakkerijlocaties en tichellanderijen.* Unpublished Research Master thesis Art History and Archaeology, Rijksuniversiteit Groningen, Faculteit der Letteren, Kenniscentrum Landschap. https://www.rug.nl/research/kenniscentrumlandschap/mscripties/remasterscritpie_j.j._wiersma_2015.pdf

14. M. Sheldrake. *Entangled Life: How Fungi Make Our Worlds, Change Our Minds and Shape Our Futures.* London: The Bodley Head, 2020.

15. A. Henare, M. Holbraad, S. Wastell. *Thinking Through Things: Theorising Artefacts Ethnographically.* London: Routledge, 2007.

16. K. Malik. 'When Monuments Fall'. *New York Review,* September 9, 2020. https://www-nybooks-com.proxy.library.uu.nl/daily/2020/09/09/when-monuments-fall/

Anton E.M. van de Ven: The challenge of living on renewable energy

1.   D.J.C. MacKay. *Sustainable Energy – without the hot air.* www.withouthotair.com, 2008.

2.   J.A. Paulos. *Innumeracy: Mathematical Illiteracy and its Consequences.* New York, NY: Farrar, Straus, and Giroux, 1988.

3.   V. Smil. *Energy and Civilization: A History.* Cambridge, MA: The MIT Press, 2017.

4.   J.W. Jevons. *The Coal Question; An Inquiry Concerning the Progress of the Nation, and the Probable Exhaustion of Our Coal Mines.* London: Macmillan, 1865, 165.

5.   J.W. de Zeeuw. 'Peat and the Dutch Golden Age: The Historical Meaning of Energy-Attainability'. *A.A.G. Bijdragen,* 21 (1978), 3-31.

6.   D.J.C. MacKay. *Sustainable Energy – without the hot air.* www.withouthotair.com, 2008.

7.   D.J.C. MacKay. *Sustainable Energy – without the hot air.* www.withouthotair.com, 2008.

8.   D.J.C. MacKay. *Sustainable Energy – without the hot air.* www.withouthotair.com, 2008.

9.   J. Sullivan. *Wind Turbines as Artistic Canvas.* https://artistsandclimatechange.com/2017/11/21/wind-turbines-as-artistic-canvas/, 2017.

10. E. Kolbert. 'The Island in the Wind'. *The New Yorker,* July 2008. https://www.newyorker.com/magazine/2008/07/07/the-island-in-the-wind

11.   J. Asafu-Adjaye, et al. *An Ecomodernist Manifesto*. http://www.ecomodern-ism.org/, 2015.

12.   IPCC. *Global Warming of* 1.5 °C, Summary for Policy Makers. https://report.ipcc.ch/sr15/pdf/sr15_spm_final.pdf, 2018.

13.   J .Lovelock. 'Op-ed'. *The Independent*, 24 May 2004.

14.   B. Spinoza. *Ethics, Demonstrated in Geometrical Order*, 1677. https://en.wikisource.org/wiki/Ethics_(Spinoza)

Filipe Freire: The earth as an observatory

1.   N. Drake. 'First ever picture of a black hole unveiled'. *National Geographic*, 10 April 2019. (https://www.nationalgeographic.com/science/2019/04/first-picture-black-hole-revealed-m87-event-horizon-telescope-astrophysics), accessed 30 December 2020.

2.   K. Thorne. *The Science of Interstellar*. New York, NY: W.W. Norton & Company, 2014.

3.   A. Cho. 'Black Holes by the Dozens Challenge Theorists'. *Science*, 370 (6517) (2020), 648.

Markha Valenta: The first assignment

1.   Malcolm X. 'Who Taught You to Hate Yourself'. Speech at Funeral of Ronald X. Stokes. Los Angeles. 5 May 1962; Malcolm X. 'Message to the Grass Roots'. Northern Negro Grass Roots Leadership Conference. Group on Advanced Leadership. King Solomon Baptist Church, Detroit. 10 November 1963.

Index of names